SOCIAL THEORY AND HUMAN BIOTECHNOLOGY

SOCIAL THEORY AND HUMAN BIOTECHNOLOGY

TIM OWEN

Nova Science Publishers, Inc.
New York

For permission to use material from this book please contact us:
Telephone 631-231-7269; Fax 631-231-8175
Web Site: http://www.novapublishers.com

LIBRARY OF CONGRESS CATALOGING-IN-PUBLICATION DATA
Owen, Tim.
Social theory and human biotechnology / author, Tim Owen.
p. cm.
Includes index.
ISBN 978-1-60692-531-7 (hardcover)
1. Social sciences--Philosophy. 2. Biotechnology. I. Title.
H61.O94 2009
303.48'3--dc22 2008053064

Published by Nova Science Publishers, Inc. ✢ New York

CONTENTS

PREFACE

The book is intended as a contribution towards metatheoretical development as part of the post-postmodern 'return to' sociological theory associated with Roger Sibeon (1996, 1999, 2004, 2007), Derek Layder (1997, 2004, 2007), Nicos Mouzelis (1991, 1993, 1995, 2007), Margaret Archer (1995, 1998) and Owen [2006a, 2006b, 2007a, 2007b] in tandem with a study of some of the sociological and ethical implications of selected examples of human biotechnology. The examples include the Human Genome Project, and related areas of interest such as reproductive biotechnology; the attempts to develop a biological sociology by writers of the 'embodied' school; and what Powell and Owen (2005) term 'the biomedical model'. The book is also intended to contribute towards 'building bridges' between post-postmodern metatheory and biological science.

ABSTRACT

The book is intended as a contribution towards metatheoretical development as part of the post-postmodern 'return to' sociological theory associated with Robert Sibeon (1996, 1999, 2004, 2007), Derek Layder (1997, 2004, 2007), Nicos Mouzelis (1991, 1993, 1995, 2007), Margaret Archer (1995, 1998) and Owen [2006a, 2006b, 2007a, 2007b] in tandem with a study of some of the sociological and ethical implications of selected examples of human biotechnology. The examples include the Human Genome Project, and related areas of interest such as reproductive biotechnology; the attempts to develop a biological sociology by writers of the 'embodied' school; and what Powell and Owen (2005) term 'the biomedical model'. The book is also intended to contribute towards 'building bridges' between post-postmodern metatheory and biological science.

The framework employed in the thesis relies upon methodological generalisations as opposed to substantive ones. It is contented that we require a 'way forward' beyond postmodern relativism, harshly environmentalist *oversocialised* accounts and *genetic fatalism* (the equation of genetic predisposition with inevitability), in the form of a modification of Roger Sibeon's ontologically-flexible, anti-reductionism. It is suggested that this modification would entail focusing upon *the biological variable* (evidence from behavioural genetics and evolutionary psychology for a, at least in part, genetic basis for some human behaviour). The mutuality between genes and environment is acknowledged in the thesis, and Matt Ridley's (1999, 2003) model of Nature via Nurture' is favoured.

The explanatory potential of the framework is demonstrated by 'applying' the meta-concepts to the study of the areas of interest. The focus here is upon the sociological aspects (in relation to the meta-concepts) of selected examples of human biotechnology and related areas, together with selected ethical concerns. It is contended that the metatheoretical framework can usefully contribute to a study of the subject area, and also towards 'preparing the ground' for further theoretical and empirical investigation, drawing upon large-scale synthesis.

ACKNOWLEDGEMENTS

I would like to record my great debts of gratitude towards the following people for their help, support, advice and inspiration in the completion of the book: Dr Roger Sibeon (Univ. Liverpool), Dr. Jason L. Powell (Univ. Liverpool), Professor Derek Layder [Univ. Leicester], Mr Ian Conolly (Nottingham Trent University), Dr Julie F. Scott (Manchester Metropolitan Univ.), Dr Steven Miles (Univ. Liverpool), Dr Dave King (Univ. Liverpool), Dr Peter Millward [Univ. Liverpool], Professor Keith Faulks [Univ. Central Lancashire], Dr Terry Hopton [Univ. Central Lancashire], Professor John Joughin [Univ. Central Lancashire], Professor Georg Wiessala [Univ. Central Lancashire] and Professor Gerard Delanty [Univ. Sussex].

Special thanks are due to my wife, Mrs Julie Ann Owen, to Mrs Sue 'young bean' Jones (Liverpool Hope University), Mark Owen SPCS Southport, and my sister Wendy and brother-in-law David Steeple.

The book is dedicated to the memory of my parents, Donald and Barbara Owen.

FOREWORD

In this impressive book Tim Owen situates his broad-ranging discussion in the context of a movement of thought aimed at a 'return to sociological theory' –or described somewhat differently- at creating post-postmodern social theory. In this sense Owen's book is an important intervention in the ongoing debate about the future of sociological knowledge –not just theory- set against worrying concerns about the about the debilitating effects that post-structuralism and postmodernism have had on social analysis and research. But Owen doesn't only engage theoretically, his views are further illuminated by reference to substantive issues around human biotechnology-specifically the Human Genome Project and reproductive technology. Given this over-arching set of concerns the book operates at several different levels.

First, he offers a concise overview of some of the main conceptual and methodological issues (along with their substantive implications), entailed in the 'return to sociological theory' debate. In this respect Owen's arguments issue from an engagement and constructive debate with the work of Sibeon, Archer, Mouzelis and Layder. These are the main intellectual supports for the kind of theory that Owen wishes to endorse and to which he adds his own distinct contribution. Owen is heavily indebted to Sibeon whose ideas he uses as a platform from which to launch swingeing critiques of examples of the 'cardinal sins' of essentialism, reductionism, reification and functional teleology, to be found in the work of writers influenced by postmodern ideas. Along with these 'sins', he detects equally worrying tendencies towards relativism and nihilism in social analysis and the knowledge it is capable of generating, not to mention the manifest deficiencies of exclusively discursive accounts of power.

Owen joins forces with the above-mentioned authors (among others) who wish to counter these tendencies and develop a more viable approach that may overcome the serious inadequacies of postmodern thought. A crucial ingredient here is the development of a flexible but 'realist' ontological framework capable of promoting constructive dialogue and genuine synthesis between otherwise antithetical analytic stances. Absolutely essential to this process is the creation of the appropriate conceptual and substantive means of dealing with central issues in social analysis such as agency-structure, macro-micro and time-space relations and robust understandings of the different forms of power in the social world. The clarity of Owen's presentation and overview of these ideas underlines the usefulness of this as a contribution in its own right. However, the discussion is further enhanced by the application of these ideas to the field of human biotechnology, which adds extra weight and sharpness to the discussion.

But Owen provides more than this. He directly contributes to the debate about the post postmodern return to sociological theory -including what he describes as 'building bridges' between the biological and social sciences. To this end he stresses that it is important to avoid collapsing the distinctions between 'biological' and 'social' components of human behaviour –a tendency he detects in some of the work in this area. To facilitate the building of bridges, Owen advances some of his own concepts. Specifically he introduces two more cardinal sins of social analysis. First, what he calls 'genetic fatalism' highlights the elicit equation between 'genetic predisposition' and 'inevitability' in genetic influences on social behaviour, an equation that underlies many discussions of this topic. Second, with 'the oversocialised gaze' Owen challenges those strongly environmentalist, social constructionist explanations –found, for example, in accounts of sexual behaviour as diverse as Gagnon and Simon on the one hand, and Foucault on the other- that dismiss genetic or partially genetic explanations.

In conjunction with these concepts, Owen wants to emphasise the importance of the 'biological variable' in social analysis and draws on evidence from evolutionary biology and behavioural science to suggest an at least in part, genetic basis for some behaviour. It is clear that Owen wants to avoid the sterile 'nature versus nurture' confrontation and to replace it with the more subtle and productive view that 'nature' operates '*via* nurture'. Support for this he finds in the work of Ridley, which stresses the mutuality and plasticity of the causal relations between genes and environment. In particular Owen agrees with Ridley's view that genes can be 'switched on' by external, environmental events, or 'free-willed behaviour'. Here Owen points out a connection with my own notion of 'psychobiography'. By this concept I originally intended to emphasise the relatively autonomous and unique psychological aspects of individual's predispositions and behaviour that interact with social influences to produce emergent effects. However, I fully concur with Owen's 'extension' of the implications of the notion of psychobiography to embrace the mutuality and plasticity of the relations between genetic and environmental influences.

All in all Owen's book is a superb attempt to overcome some of the more negative effects that the recent, so called 'cultural turn', has had on social analysis. He opens up new ground and endeavours to provide some of the conceptual means whereby more productive debate might ensue, and thus help precipitate a reinvigoration of sociological analysis.

Professor Derek Layder, University of Leicester, UK, 2008.

Chapter One

Introduction

It is the intention here to contribute towards theoretical development as part of the post-postmodern 'return to' sociological theory and method associated with Sibeon (1996, 1999, 2004, 2007), Layder (1997, 2004, 2007), Mouzelis (1991, 1993, 1995, 2007), Archer (1988, 1995, 1998) and Owen [2006a, 2006b, 2007a, 2007b], in tandem with the study of the *selected* sociological (in relation to the meta-concepts of the framework) and *selected* ethical implications of *selected* examples of human biotechnology, such as the Human Genome Project and related areas of biosocial (involving both social science and biology) interest such as the techniques of reproductive human biotechnology identified by Lynn (2001) as having eugenic potential; attempts by those of the 'embodied' school such as Shilling (1993) and Newton (2003) to develop a *biological sociology*; and the 'bio-medical' model of theorising (Powell and Owen, 2005). It is envisaged that the metatheoretical framework will also contribute towards the 'building' of 'bridges' between sociological theory and biological science. The 'sensitising' framework relies upon methodological generalisations as opposed to substantive generalisations. It is the contention here that we require a 'way forward' beyond, on the one hand, the anti-foundational relativism of the postmodern and poststructuralist approaches, and on the other, *oversocialised* (harshly 'environmentalist') accounts and *genetic fatalism* (the equation of genetic predisposition with inevitability) in the form of a synthetic 'modification' of Sibeon's (ibid) original, anti-relativistic, ontologically-flexible *anti-reductionism.*

In what follows, it is suggested that a modification of Sibeon's (ibid) framework which would entail focusing upon the *biological variable* (evidence from evolutionary psychology and behavioural genetics for a, at least in part, genetic basis for some human behaviour) has considerable explanatory potential in relation to the study of *selected* examples of human biotechnology and *selected* related biosocial areas. Here, we acknowledge the elegant *mutuality* between genes and environment and favour the evolutionary psychologist, Matt Ridley's (1999, 2003) model of *'Nature via Nurture'*. Meta-concepts from the post-postmodern framework are 'applied' to the study of the previously outlined areas of interest. The focus here is upon the *selected* sociological aspects (in relation to the framework's meta-concepts) and *selected* ethical concerns regarding selected examples of human biotechnology. There is no intention of reviewing the vast literature on *bio-ethics* or of 'taking sides' in ethical debates. Regarding ethical concerns, the emphasis is upon eugenic concerns, and this reflects an interest/concern on the behalf of the author. It is also felt that it is absolutely necessary to examine ethical concerns in relation to human biotechnology because the two subjects are so closely linked together in the minds of members of the public and the 'academic community'. To ignore such issues would render the work less credible. Issues of *risk* and *governance* are certainly mentioned in the work, and the relevant literature covered, but the intention here is *not* to explore such avenues at depth.

It will be noted that the work consists of five integrated and interlocking Chapters and Concluding Observations. This Introduction is the first of the Chapters in which the aims and objectives have been outlined. We now move on to examine the contents of each Chapter and how they 'fit' together. It will be recalled that the ultimate intention is to develop a meta-theoretical framework which will contribute towards post-postmodern theoretical development, towards the 'building' of 'bridges' between sociological theory and biological science, and also prove its worth in terms of explanatory potential in relation to *selected* examples of human biotechnology and related areas. We need to examine the individual stages of the process, as they unfold through the Chapters.

Chapter Two ('Transitions in Social Theory') reveals how there has been a mounting reaction in contemporary sociological theory against the nihilism and paralysis of anti-foundational, relativistic postmodern and poststructuralist approaches. Sibeon's (1996, 1999, 2004, 2007) *anti-reductionist* framework is suggested as a 'way forward' beyond such relativism, but it is observed that it will be necessary to 'modify' the metatheoretical framework so that it includes 'new' terms such as *the oversocialised gaze*, *genetic fatalism* and *the biological variable* in order to conceptualise biotechnological issues and to contribute towards the study of human biotechnology/related biosocial issues. The Chapter shows how the framework (like Sibeon's original) is an example of *metatheory*. Metatheory is designed to equip us with a 'general sense of the kinds of things that exist in the social world, and with ways of thinking about the question of how we might 'know' that world' (Sibeon, 2004: 13). It entails a flexible ontology, avoiding the relativism of the 'cultural turn', consisting of methodological generalisations rather than substantive ones. Here we define and cover the 'Cardinal Sins' of *reductionism; reification; essentialism;* and *functional teleology* out of which Sibeon's original framework, which focuses upon *agency-structure, micro-macro* and *time-space* (and is anti-relativistic) arose. We examine the merits of a metatheoretical approach in sharp contrast to that of anti-foundational relativism (the section on 'Metatheory and the Problems of Relativism'). Lyotard's (1986) post-modern approach is our main focus as it could be argued that the French writer did more than any other thinker to disseminate postmodern positions. It is decided that the deficits of the 'cultural turn' outweigh the benefits of social analysis and it is reiterated that we require a post-postmodern approach of the sort employed in the 'new' framework and by Owen (2006a, 2006b, 2007a, 2007b) in order to study human biotechnology and related areas. Next, we examine the controversial Human Genome Project in relation to the 'modified' version of Foucauldian power employed by Sibeon (2004). This is in order to demonstrate early on in the work, the usefulness of a meta-concept in relation to a selected example of biotechnology. It is emphasised that the 'real' application of meta-concepts to the area of study takes place in Chapter Five. The 'background' to the Genome Project is discussed in greater detail in Chapter Four. It is made clear in Chapter Two that Sibeon's (ibid) use of Foucauldian power, which regards *systemic* and *agentic* powers as autonomous, and acknowledges the *dialectical* relationship between *systemic* and *relational* powers can contribute towards conceptualising power-relations pertaining to the Genome Project. The Chapter then considers issues pertaining to 'Agency and Social Actors' (section of this title), and the importance of the terms *agency-structure* and *micro-macro* is established in relation to how they refer to 'differing dimensions of social reality' (Sibeon, ibid: 34). A section on the work of Margaret Archer, Derek Layder and Nicos Mouzelis follows in which their considerable scholarly contributions towards social theory are discussed, and we consider how the authors conceptualise *agency-structure, micro-macro* and deal with the question of whether to employ *dualism* or *duality of structure*. It is decided that Layder's (1997) concepts of *social domains* is the more closely related approach to Sibeon's original framework. Having examined the concepts and illegitimate forms of reasoning ('Cardinal Sins' of *essentialism, reification, reductionism* and *functional teleology)* which the author's original anti-reductionist framework focuses upon and arises as a critique out of respectively; the problems of the 'cultural turn'; *power* in relation to the Genome Project (in an early 'taster' of the framework's potential); issues pertaining to *agency-structure* and *micro macro;* the contributions of three major theorists to the 'transitions' that have prevailed in contemporary theory, we move forward to Chapter Three. We must keep in mind here the idea that it might be necessary to 'modify' Sibeon's original framework to include a focus upon the 'new' terms; *the biological variable, genetic fatalism* and *the oversocialised gaze* in order to progress.

Chapter Three ('A Post-Postmodern Metatheoretical Framework?') takes the suggestion to modify Sibeon's framework considered in Chapter Two, and 'weighs up' the evidence from evolutionary psychology, behavioural genetics and biological science for, at least in part, biological causality in relation to *selected* examples of human behaviour. This is absolutely necessary in order to properly justify the inclusion of 'new' concepts such as the *biological variable, the oversocialised gaze* and *genetic fatalism*. Firstly, the framework is briefly codified in order that the reader is 'kept on track', so to speak. The section 'Building Bridges' takes the reader through an examination of the work of several authors who have attempted to straddle the biology/sociology divide. In addition to recent attempts by Owen (2006a; 2006b, 2007a, 2007b) to apply the meta-concepts of this framework to social gerontology, biotechnology and criminal behaviour respectively; attempts by Quilley and Loyal (2005) which employ Elias's figurational theories in an attempt to synthesise social and biological sciences; and Freese et al's (2003) work on '*biophobia*' (hostile reactions on the behalf of social scientists towards genetic explication – similar to the notion of *the oversocialised gaze* employed here in the sense of antipathy towards all but 'environmentalist' accounts) are mentioned. The main focus, however, is upon the more well-known attempts by sociologists of the 'embodied' or 'material corporeal' school to 'build bridges' between social/life sciences. Drawing largely upon the work of Ridley (1999, 2003), we show how genetic predisposition (Ridley uses the phrase 'determinism' in the same sense as predisposition, 1999: 307) need *not* equate to inevitability because of the *mutuality* between genes and environment. It is suggested here that an acknowledgement of Ridley's (ibid) '*Nature via Nurture'* model is a possible 'way forward' beyond Shilling's (1993) 'starting point' for a *biological sociology*, by acknowledging what material corporeal sociologists will not – 'genes do influence behaviour' but they can be 'switched on' by free-willed, external, environmental stimuli (Ridley, ibid). We next examine the sources of evidence for biological causality in respect to selected human behaviours, focusing largely upon evolutionary psychology. The section, 'Evolutionary Psychology and 'Nature via Nurture'' considers detailed, in-depth criticisms of evolutionary psychology by those such as David (2002); Rose and Rose (2000) and others, and the contrary evidence in favour of the approach from Curry (2003); Fodor (1983); Pinker (1995; 1999); Tooby and De Vore (1987); Daly and Wilson (1998) and others. It is concluded that we *can* acknowledge the evidence from evolutionary psychology and related disciplines to the effect that it *does* seem likely that human beings posses *instincts* in the sense of unlearned patterns of behaviour, as, for example, Hamer and Copeland's (1999) work on sex-drive appears to confirm. Following this, we examine the need to incorporate dualism in the sense of a dualistic conception of biology and 'the social', alongside Sibeon's dualistic conception of *agency-structure*. The section 'Dualism and Duality Revisited' further explores issues around dualism/duality which were first briefly encountered in Chapter Two's section on the work of Derek Layder, Nicos Mouzelis and Margaret Archer. After deciding that a *dualistic* conception of biology and 'the social' is an effective way to avoid what Archer (1995: 96) calls *'central conflation'* (the author refers specifically to Giddensian tendencies to collapse the distinction between agency and structure), we move on to the final section of Chapter Three – 'Incorporating the Biological Variable'. Here, we look more closely at the evidence for biological (or partially biological) causality in relation to *selected*, specific human behaviours such as *sexuality* (Hamer and Copeland, 1999); *human reactions to stress* (Ridley, 2003); and *language* (Enard et al, 2002). We make the decision, in the face of the cogent evidence for a biological component to the above behaviours to *incorporate* the *'biological variable'* into the meta-framework alongside Sibeon's anti-relativistic, ontologically-flexible notions of *agency-structure, micro-macro, time-space, power* and *dualism* (and, of course, the 'cardinal sins' of *essentialism, reductionism, functional teleology* and *reification*). If we can incorporate *the biological variable*, then the related 'new' meta-concepts of the *oversocialised gaze* and *genetic fatalism* must be incorporated too. It is noted at the end of Chapter Three that having made such a decision, based upon a thorough review of the evidence, we can go on to codify the 'new' expanded framework and 'apply' the meta-concepts to the area of interest/study in order to demonstrate explanatory potential in Chapter Five. However, before we are able to do so, it is absolutely necessary to examine general ethical concerns (with an emphasis upon eugenic potential) in relation to selected examples of human biotechnology, including the Genome Project itself, and related bio-social areas of concern such as 'the bio-medical' form of theorising (Powell and Owen, 2005).

In Chapter Four ('Human Biotechnology: Issues and Implications') we are concerned more than anything, in the main, with providing the 'background' to the Human Genome Project as perhaps the most controversial and publicly 'visible' bio-medical research programme and in effect a 'symbol' of human biotechnology in the minds of many, and the 'background' to the selected techniques of reproductive human biotechnology which Lynn (2001) associates with greatest ethical controversies and eugenic concerns. This is in order to provide the reader with sufficient information about the areas of study, and to 'prepare the ground' for the 'application' of meta-concepts to the Genome Project, reproductive technologies etc in the following Chapter Five. The first section 'The Human Genome Project' describes the 'history' of the huge project, including criticisms from those such as Wilkie (1994), Opitz (2000), Lewontin (1992) etc along ethical lines. Material from 'sociologists of science' such as Bucchi (2004) is considered, and the Project is conceptualised as exemplifying 'a number of ongoing patterns of change in contemporary science' (ibid: 134). This is in terms of being created for 'political purposes', which is 'typical' of post-war research policy and 'developed in coincidence with profound changes in the organisation of scientific research'; a 'post academic science' (ibid). This 'post-academic' science is characterised by the intersection of disciplines, the heterogeneity of the actors involved, and by reflexivity and ideas pertaining to social accountability (Gibbons et al, 1994). Additionally, the Project is viewed by Bucchi (ibid) as a new, transformed kind of 'Big Science' which has ousted the old 'military-industrial complex' replacing it with a new 'academic-industrial-governmental' complex. Next, we very briefly consider issues of *risk* and *governance* in relation to human biotechnology, but the intention here is *not* to look at them in depth. It is contended here that ethical eugenic issues are more closely associated with human biotechnology than are *risk* and *governance*. These are legitimate areas of enquiry, and we briefly examine the relevant literature on both subjects, but the focus is upon ethical concerns mainly pertaining to eugenics and, as we see in relation to later material in the Chapter on 'bio-medical' theorising, on sociological concerns pertaining to the meta-concepts of the framework. Lynn's suggestion (2001) that there are six techniques of human biotechnology that, in his view, have been used to promote eugenics, or that have the potential to promote eugenics in the future is fully explored next. Lynn's extremely controversial views on geographical variations in intelligence are neither endorsed nor discussed here as they are irrelevant to the purposes of the book. We look at the areas of artificial insemination; egg donation; pre-natal diagnosis; pregnancy termination of 'defective' fetuses etc and their eugenic potential. To reiterate, it is felt that if such issues were ignored, the work would lack credibility because ethical concerns of this nature are timely, topical and are linked closely in human imagination with rapid advances in human biotechnology, 'the new genetics' etc. The Chapter's final section is on 'Bio-Medical Discourse'.

It is contended that, having provided a general overview of selected sociological and ethical implications regarding selected examples of human biotechnology and the Human Genome Project itself, it is necessary to examine what Powell and Owen (2005) call 'the B*io-Medical Model'*. We have already noted that it is necessary to avoid *oversocialised* perspectives which deny genetic causality altogether, such as Gagnon and Simon's (1973) work on 'learnt' sexuality. However, it is equally important to keep in mind Ridley's (1999, 2003) conception of 'nature via nurture' in which genes *and* environment play a *mutual* part in human behaviours. Therefore, we also need to avoid *essentialist, reductionist*, totalised discourses such as 'the bio-medical model' of theorising which denies environmental causality too. We must acknowledge that there is *plasticity* between 'nature' and 'the social'; and that genes '*take their cue'* from nature. With a critique of the 'bio-medical' approach, we have paved the way, having considered the 'transitions in social theory' in Chapter Two; the evidence for genetic causality in Chapter Three; and the selected sociological/ethical implications of selected examples of human biotechnology and the Human Genome Project in Chapter Four, for an 'application' of the meta-concepts of the 'new' synthetic framework to the study of human biotechnology in Chapter Five.

Now, with the 'ground prepared', we may 'apply' these meta-concepts to the area of study. It will be recalled that, at certain points, such as Chapter Two's consideration of the effectiveness of modified notions of Foucauldian *power* in explanations of power-relations pertaining to the Genome Project; and in Chapter Four's 'application' of certain meta-concepts (agency, functional teleology etc) to the 'bio-medical' approach, we have already 'applied' concepts to the area of study. However, this has not been done in the systematic depth that we encounter in Chapter Five. The fifth Chapter opens with a codification of the framework, and takes the reader through definitions of each term. Next, the task is to demonstrate the *explanatory potential* of the individual meta-concepts of the framework. Each meta-concept is given its own heading and the term 'applied' to selected areas of human biotechnology/areas of biosocial concern. For example *reification*, the 'cardinal sin' involving the illegitimate attribution of *agency* to entities which are not actors is identified in the work of Wilkie (1994: 187) when he claims that the Human Genome Project is typical of institutions which, 'have a history and their own inertia and 'will to live''. It is present in the way Callon (an actor-network theorist) conceptualises 'scallops' and the Chernobyl nuclear plant as 'actors' (1986: 204). Pickering's *posthumanist* 'decentred' variant upon the social-constructivist sociology of science engages in reification when attributing *agency* to non-material objects (both *actor-network* and *posthumanist* theories are regarded as, to an extent, attempts to make links between social and life sciences and thus relevant to our concerns]. After Chapter Five's detailed 'application' of all metatheoretical concepts to the areas of study it is necessary to offer some final, concluding observations. These include a detailed summary of what an 'application' of each concept 'revealed'. In other words, how useful the meta-concepts were in terms of explanatory potential in relation to the areas of investigation.

Chapter Two

TRANSITIONS IN SOCIAL THEORY

There appears to be a mounting reaction in contemporary theory against the 'cultural turn' and the extreme relativism of postmodern and post-structuralist theory. To an extent, it could be argued that there has been a 'return to' sociological theory and method reflected in the work of Mouzelis (1991, 1993a, 1996, 2007), McLennan (1995), Holmwood (1996), Stones (1996), Sibeon (1996, 1997a, 1997b, 1999, 2001, 2004, 2007), Layder (1984, 1994, 1997, 2007), Archer (1982, 1988, 1995, 1996, 1998, 2000), and Owen [2006a, 2006b, 2007a, 2007b]. This so-called 'return to' sociology has been the 'accumulation of relatively separate intellectual moves that are a blend of renewed interest in classical sociology and in perennial explanatory problems, together with theoretical reflection arising from critical engagement with comparatively recent perspectives that range from neo-functionalism to actor-network theory' (Sibeon 2001: 1).

While theorists such as Gellner (1993) view postmodernity (and, in particular, the Lyotardian version) as practically worthless, here the view is that aspects of it are relevant to the development of sociological theory and method (Lemert 1993; Sibeon ibid) in so far as it rejects 'modern' essentialist and reductionist theories. However, despite some useful concepts pertaining to contingency, one concurs with Holmwood's (ibid) critique of postmodern theory. If, as Lyotard (1986) claims, no coherent theory is adequate, how are we to decide which incoherent postmodern theories *are* adequate? Here a 'way forward' is suggested in the form of an anti-reductionist, post-postmodern sociology. We need a form of sociological 'realism' though not in the sense of reductive and reified material evidence. While recognising the structural qualities of social systems, an anti-reductionist sociology would recognise that even if social reality *is* socially constructed, 'there may still be a socially constructed reality 'out there'' (Sibeon 1997a: 3).

Here, I cautiously suggest a modified version of Sibeon's *anti-reductionist* framework consisting of methodological generalisations as distinct from substantive generalisations, for the purpose of studying human biotechnology. The framework is designed to stimulate the development of non-reductive theories and empirical investigations. Above all, the framework is a *synthesis*. Its methodological reach owes much to several theoretical schools mentioned in the introduction. Sibeon's original *anti-reductionist sociology* arises out of a critique of four illegitimate forms of theoretical reasoning; these are *reductionism, essentialism, reification,* and *functional teleology*. A *reductionist* theory is a theory that attempts to explain social 'reality' in terms of a single, unifying principle such as 'patriarchy'

(Hindess 1986, 1988). *Essentialism*, 'presupposes in a prioristic fashion a necessary 'unitariness' or 'homogeneity of social phenomena' (Sibeon 1997b: 1). *Reification,* 'involves the illegitimate attribution of agency to entities that are not actors or agents' (ibid). *Functional Teleology*, refers to 'illicit attempts to explain the causes of social phenomena in terms of their effects' (Betts 1986: 51). My modification of Sibeon's original framework arises out of a critique of the following, additional illegitimate forms of reasoning; these are *Genetic Fatalism* and the *Oversocialised Gaze. Genetic Fatalism* refers to a widespread tendency within contemporary social science to equate genetic determinism with inevitability. *The Oversocialised Gaze* refers to 'Environmentalist', sociological accounts characterised by a strong antipathy towards genetic or even partially genetic explanations.

I have also modified Sibeon's original framework so that it entails focusing upon the *biological variable,* in addition to *agency-structure*, *micro-macro, time-space* and *power* as core concepts. The *'biological variable'* refers to the evidence from over twenty five years of research into behavioural genetics for a, at least in part, biological basis for some human behaviour. This emphasis upon biological variables reflects the influence of insights from recent work which attempts to cross the biology-society divide associated with Newton (2003), Williams (1998, 2003), Bury (1997), Benton (1991, 1994) and insights from the evolutionary perspectives of Daly and Wilson (1998), Tooby and De Vore (1987), Curry (2003) and others. Here, it might be useful to describe a number of the core concepts that originally formed part of Sibeon's *anti-reductionist sociology,* and are now included within the 'new framework'. As stated in the Introduction, the 'new framework' has been used to contribute towards a study of human biotechnology (Owen, 2006b), a post- Foucauldian sociology of aging (Owen, 2006a), towards an anti-reductionist approach to aging (Powell and Owen, 2005), towards a study of 'trust' and professional power [Owen and Powell, 2006], and in the context of approaches to crime [Owen, 2007b]. Here the emphasis is upon a more detailed examination of (selected examples) of human biotechnology and the Human Genome Project. In what follows it is essential to examine Sibeon's original framework in detail.

FOUR 'CARDINAL SINS'

Here it is necessary to discuss and shed light upon four 'cardinal sins'; illegitimate forms of theoretical reasoning that are singled out for particular criticism in Sibeon's original framework. Indeed, the original framework arose out of a critique of the terms. *Reductionism, essentialism, reification* and *functional teleology* are also included in my modified framework, alongside additional forms of illegitimate reasoning such as *genetic fatalism* and the *oversocialised gaze* (Owen, 2006a, 2006b)

According to Sibeon (2004: 2) whether or not the term 'crisis' has been overused in relation to sociological theory and method, it can be argued that many of sociology's problems of theory and 'explanatory failure' have been associated with a tendency to draw upon the four forms of illegitimate reasoning that are *reductionism, essentialism, reification* and *functional teleology.* Nevertheless, there are reasons for optimism. Despite undoubted difficulties presented by the legacy of postmodernism in terms of *relativism,* there is, 'a major drive towards a post-postmodern 'return to' sociological theory and method (McLennan,

1995)' which 'seems both feasible and desirable (Archer, 1995, 2000; Mouzelis, 1995; Stones, 1996; Ritzer, 2000; Turner and Rojek, 2001)' (Sibeon ibid). What follows is an attempt to define the original four 'cardinal sins'.

Reductionism

Sibeon (ibid) suggests that *reductionist* theories are ones which attempt to reduce 'the complexities of social life to a single, unifying principle of explanation or analytical prime mover (Hindess 1986a, 1988) such as 'the interests of capitalism', 'patriarchy' 'rational choice', 'the risk society', 'trust', 'the information society', 'globalisation', or whatever'. The author also cites examples in accounts of government and the policy process, theories of the state etc (theories such as pluralism, elitism, or Marxism) which are reductionist in the sense that each of them is 'predicted on the view that government and public policy can be reduced to a single substantive principle of explanation' (ibid). Reductionist, 'general' theories are 'ontologically inflexible', resting upon 'a priori assumptions about the nature of the state and of society in regard to factors affecting the distribution of power' (ibid), and in respect of the nature of political/policy dynamics. Sibeon cogently suggests that, 'a more adequate and empirically sensitive form of enquiry' would recognise that certain policy sectors such as education, health and foreign policy, for example, may be 'pluralist', whereas others are, 'dominated by policy networks that have an elitist or corporatist form'; and it is possible that 'power distributions and policy dynamics within each sector may shift over time', or they may vary spatially (Sibeon, ibid). Sibeon goes on to suggest, using the present example, that a better understanding of the 'complexities of politics, power and public policy' is likely to be achieved by use of 'non-reductionist, ontologically flexible methods of investigation' such as 'policy network analysis (Rhodes, 1997; Marsh, 1998; Marsh and Smith, 2000) which is a contemporary political science approach' (Sibeon ibid: 3). There are, according to the author, three general aspects of his original *anti-reductionism.* Firstly, 'to reject a priori theoretical commitment to analytical prime movers is not to say there are never situations where a very small number of factors' may have causal primacy (Sibeon, ibid). This should be treated as an 'empirical question' rather than as something, in advance of empirical investigation, that is theoretically predetermined 'on the basis of some reductionist social theory'. It is not suggested that substantive theories should, like 'grand theories' attempt to encompass all the possible factors that may shape the phenomena under investigation. Every substantive theory is partial in terms of the type/number of phenomena to be explained. Secondly, Sibeon distinguishes between 'non-reductive multi-factorial explanation', in other words, explanations which draw upon a cluster of variables, and 'compounded reductionism' (ibid). The latter involves attempts to synthesise/combine two or more reductionisms. The author cites Dominelli's (1997) attempts to synthesise reductionist theories of capitalism and patriarchy, arguing that this results in contradiction and 'explanatory failure'. Thirdly, Sibeon cautions against 'deferred reductionism'. This 'takes an Althusserian-like form' where reductionist explanation is postponed or deferred in social analysis, until the last instance (ibid). An example, the author suggests, lies in Farganis's feminist critique of postmodern theory, in which a straightforward or obvious reductionism is replaced by a seemingly non-reductionist and multi-factorial approach, which turns out to have a reductionist theory at its base. Another example offered by Sibeon is Harvey's attempts to blend Marxism with postmodernism despite references to contingency. Harvey associates the animation of the social with a 'prime mover', namely modes of production (1989: 107), (Sibeon, ibid).

Essentialism

Essentialism is, a 'form of theorising that in a prioristic fashion presupposes a unity or homogeneity of social phenomena', such as the law or other social institutions, or taxonomic collectivities such as 'women', 'men', 'the middle class' or 'white people' (Sibeon, ibid: 4). The author acknowledges the inescapability of a certain degree of idealisation of phenomena in order to be able to refer to the phenomena at all. However, we should not, 'falsely essentialise them or theoretically ascribe to them more homogeneity than they actually possess' (Sibeon, ibid). Essentialist reasoning 'does not regard the degree of homogeneity or heterogeneity of social phenomena as an empirical variable for investigation', but rather 'presupposes on theoretical or political grounds a 'necessary' unitariness of the phenomena' under investigation (ibid). The author cites Albrow's (1996: 91-2, 94-5) criticisms of essentialist notions of globalisation as highly pertinent. Sibeon's view is that 'essentialist thinking' is linked to or is 'a corollary of a reductionist theory', and like reductionism, essentialism can take on a 'disguised' or 'last instance' form as, for example, when 'women' or 'the working class' are described as social categories that, 'though perhaps acknowledged to be in some respects internally divided and cross-related to other categories', should be regarded as 'ultimately possessing a 'primary, overriding commonality' that transcends all other memberships (categories) or affiliations (ibid). The author argues that if unitariness is a feature of any social phenomenon, it is to be regarded as a contingent and emergent – and perhaps also a temporary outcome of social processes rather than a necessary effect of social totality.

Reification

Reification is 'the illicit attribution of agency to entities that are not actors or agents' (ibid). Drawing upon Harre's (1981) concept of agency and Hindess's (1988: 45) 'minimal concept of an actor', Sibeon argues that an *actor* or agent, 'is an entity that in principle has the means of formulating, taking, and acting upon decisions' (ibid). This is a non-reified definition of a social agent or actor. On this basis, there are only two types of actors: individual human actors; and 'social actors' (Hindess, 1986a: 115) or 'supra-individuals' as Harre (1981: 141) terms them. According to Sibeon, the latter are organisations (such as government departments, professional associations); committees, such as the Cabinet or micro-groups such as households. The author offers examples of *non-actors*, these being 'entities that cannot express agency', in other words, entities that cannot formulate or act upon decisions – such as 'society', 'the state', and taxonomic collectivities such as social classes, 'the British people', 'young people', and so on (ibid: 5). This non-reified conception of agency challenges the view that, for example, men are an entity (an actor) that can take collective action to remedy gender 'issues'. In so far as agency is a key variable in the production/reproduction of structure, it is incontrovertible that only actors as defined in the original framework can have causal responsibility for social conditions (including inequality), and only actors are capable of formulating and carrying out actions that reproduce or alter those conditions. As Sibeon makes clear, reification, like the other 'cardinal sins' identified in his original framework, is not confined to particular theories/paradigms. Rather, it occurs almost everywhere across the theoretical landscape. He provides a few examples to illustrate this point.

The author cites the work of Touraine on a sociology of action. Sibeon disputes Delanty's (1999: 122-44) description of Touraine as a theorist committed to the 'return of agency' to social theory. He feels that Delanty's overall assessment of Touraine is 'rather too generous' as the latter's 'conception of agency and social action is marred by reification' (ibid). For example, Sibeon shows how, in Touraine's *The Voice and the Eye*, 'society' is regarded as an actor (1981: 31, 59) and social movements are regarded too as 'actors', as are social classes. In recent work, Touraine's (1995) conception of agency and social action continues to engage in reification, according to Sibeon (ibid).

In a similar way, Habermas's central theories exhibit tendencies towards reification. For example, Habermas's (1987: 159-60) attribution of agency to social systems is singled out here; specifically the idea that social systems regulate their exchanges with their social and natural environments by means of co-ordinated interventions into the social world. Sibeon also identifies strong tendencies towards reification in the following: Luhman's (1982: 265) attribution of agency to what he has termed 'autopoietic social systems'; John Law's (1991b: 173-4) blending of post-structuralism with actor network theories (formerly translation theories), which leads the latter to the view that 'an agent is a structured set of relations'; Foucault's claims to the ends that discourses are agents (Danaher et al, 2000: 33), and the 'posthuman agency' of Callon (1986), Latour (1988) and Pickering's (2001) use of post-structuralism and actor network theories (Sibeon, ibid: 6). Pickering and other actor network theorists appear to suggest that the 'natural' material world and objects such as stones can be said to display agency. Jones (1996: 296) rejects such a conception of agency as an 'obscure, hollow metaphysics'.

Functional Teleology

According to Sibeon (2004: 6), this is 'an invalid form of analysis involving attempts to explain the causes of social phenomena in terms of their effects', where 'effects' refers to outcomes or consequences viewed as performances of 'functions'. The author, requesting that we bear in mind his definition of non-reified actor/agent here, argues that in the absence of international planning by actors somewhere/sometime, it is a 'teleological fallacy' to attempt to explain the causes of phenomena in terms of their effects (Betts, 1986: 51) (Sibeon: ibid). Reductionist theorists may begin with what Sibeon calls, a 'current social or cultural item' (a social practice, or a law, or a public policy etc), and then endeavour to 'work backwards' claiming without any evidence of intentional planning by actors, that the item came into being because it 'accorded with the interests' of a taxonomic collectivity such as 'white people' or 'men' (ibid: 7). Related to a neglect of agency, teleology is flawed by a problem of logic in so far as the factors that bring a social or cultural item into being must predate the existence of the item already existing. Sibeon cites Durkheim, in some of his theoretical work as 'guilty' of engaging in functional teleology, although he shows awareness in his methodological work, of 'the importance of separating causal explanation from functionalist explanation'

(Durkheim, 1982: 90, 95) (Sibeon, ibid: 7). Famously, Durkheim argued for the need to distinguish between the *causes* of something and the *functions* of something. In Sibeon's view, illegitimate functionalist teleology conflates causal and functionalist explanation and is an attempt to explain the cause of phenomena in terms of the item's 'functions', that is to say, in terms of the item's consequences/effects. Later in the Chapter, concerning an application of the framework to the study of human biotechnology, we encounter an example of *functional teleology* in the work of Lynn (2001). Lynn is concerned with whether such medical procedures (pre-natal diagnosis, AID etc) are eugenic measures, and whether this can be determined by considering the intentions of the physicians carrying out the procedures, rather than by their effects. Lynn (ibid: 24), as is suggested by Owen (2006b), is arguably wrong in suggesting that the criterion of the intention of the physicians carrying out the procedures is unsatisfactory because it is difficult to know what their intentions are, especially because no physicians are going to admit publicly that they are endeavouring to promote eugenics. To analyse the issue in terms of the effects of the procedures would be to engage in *functional teleology*, an illegitimate form of theoretical reasoning.

One of the major themes of this book is that Sibeon's original framework, subject to modifications which are outlined in greater detail in the following Chapter, can be used to look closely at the sociological implications of selected examples of biotechnology. The intention in what follows is to demonstrate that the framework is a form of what Sibeon calls 'metatheory', and to discuss in some depth issues such as agency and social actors. We will be drawing upon the work of leading sociological theorists such as Margaret Archer, Nicos Mouzelis and Derek Layder who are part of the transition in social theory towards a post-postmodern 'return to' sociology.

Metatheory and Problems of Relativism

Here, the task is to outline the role that metatheory plays within social theory in general. The rationale for employing metatheory is discussed, and there is a review of controversies surrounding the problem of relativism and postmodern/poststructuralist rejection of theoretical foundations. The framework developed in this book is very much an example of 'sensitising' metatheory, and a development of Sibeon's original *anti-reductionism.*

As Sibeon (2004: 13) suggests, metatheorists are largely 'concerned with ontological questions', including ones such as, 'What is 'society'? What type of things exist in the social world?' If actors and agents exist, what sort of things are they? 'Does it make sense to employ a stratified social ontology that refers to micro and macro spheres or 'levels' of society, or is micro-macro – as Foucault, Elias and Giddens claim – a spurious and misleading distinction?' (ibid). The author also poses the question, 'And when we turn to epistemology, it is important to ask: how is reliable knowledge acquired?' (ibid). The question is also posed as to whether 'lay' actors' self-experience is a 'better guide' to 'social reality' than the 'objective' (in theory) perspective of social scientists? Perhaps the fundamental question, as Sibeon suggests, is whether objective, social reality exists 'out there', independent of our conception. Often the term *sensitising theory* is used in social theory rather than *metatheory*. According to Sibeon, metatheory can and should inform the development of substantive theory, but the types are distinct. Metatheory, such as the framework employed in this book, is designed to equip us with 'a general sense of the kinds of things that exist in the social world, and with ways of thinking about the question of how we might 'know' that world' (ibid). In contrast, substantive theories 'aim to generate new empirical information about the social world' (ibid). Mouzelis (1993a: 684) makes a distinction between substantive generalisations and methodological generalisations. He notes, as Sibeon (ibid) points out, that 'their aim is less to tell us things we do not know about the social than to provide us with conceptual tools for asking interesting questions and preparing the ground for the empirical investigation of the social world'. Therefore, the role of any metatheoretical concept (such as the framework employed in this book) is to generate at the meta-level, conceptual tools that inform the development of concepts, substantive theories and explanatory schemes and that underpin the design of empirical studies.

As Sibeon (ibid: 14) notes, 'not all researchers and theorists are happy with the idea of metatheory'. Postmodernists such as Lyotard (1986) tend to reject metatheory on the grounds that it is a 'grand narrative'. Owen (2006b) suggests, Lyotard (celebrated in some circles as the postmodern theorist *par excellence*) is an example of a theorist of the 'cultural turn'. Owen raises the question of whether Lyotard rejects foundationalism from which theory can be generated; and whether he provides an acceptable postmodern epistemology and a viable postmodern theory. The most basic criticism of Lyotard's (1973, 1986, 1986-7) relativistic position is that he never applies it to himself, to his own theories and conceptual frameworks. Lyotard is open, that is to say, to *the self-referential objection* (Blackledge and Hunt, 1993) which posits that, if all theories are the product of a particular situation, then so too is *that* theory, and it therefore has no universal validity. To put this another way, if truth and falsity do not exist in an absolute sense, then Lyotard's (1986) thesis about the relativity of all knowledge cannot be 'true' in this sense. His thesis, however, invalidates itself; Lyotard is arguably hoist on his own petard. In arguing the way he does, Lyotard is surely employing the very criteria of truth and validity which he claims are culturally relative. He informs his reader that norms of truth change, and that ideas of a 'good explanation' are matters of 'social convention' (Lyotard 1986: 80). Yet he himself is using such conventions in saying this. He is, in a sense, employing reason to try to prove the inadequacy of reason; claiming to provide a universally valid and 'true' explanation of why there is no such thing as a universally valid and 'true' explanation. Lyotard's endeavour is therefore highly problematic and contradictory. As Gurnah and Scott (1992: 148) observe, postmodern theorising rests upon, 'a grand narrative.... It thus reproduces precisely those totalising characteristics of which at the level of intellectual critique it is most critical'. Put simply, the postmodern statement that there can be no general theory 'is itself a general theory!' (Sibeon, 996: 14). There are those, however, who regard postmodernism as a 'dead' theory [Ritzer and Ryan, 2007].

Sibeon (2004: 14) makes the point that 'some meta- or sensitising theories formulate 'large' generalisations pertaining to common social processes that may be found in a wide variety of social settings'. He gives the example of Giddens's (1989: 295) structuration theory, which is a reasonably well-known form of metatheory, consisting of postulates that are 'intended to apply over the whole range of human social activity, in any and every context of action'. Such *meta*theoretical generalisations of this type are not the same thing as 'grand' generalisations 'associated with reductionist substantive theories such as Marxism, rational choice theory, and radical feminism' (Sibeon, ibid), theories which attract the criticism of postmodernists such as Nicholson and Seidman (1995: 7). In contrast, Giddens's structuration theory, according to Sibeon (ibid) is an example of metatheory 'that does not invoke the reductionism associated with 'grand' substantive theories' such as radical feminism (which reduces the complexity of social relations to the idea of patriarchal causality); 'structuration theory is ontologically flexible (Cohen, 1987: 279-80, 285, 289, 291), a term that refers to metatheory of a kind' which makes possible the development of a range of competing substantive theories, and for 'open-ended' empirical investigation of phenomena.

As we have observed with Lyotard, postmodernists deny the validity of sociological (or any form of disciplinary discourse) discourse. So, if one wishes to employ a metatheoretical framework such as the one employed in this study, it is essential to address the emergence of relativistic postmodern theory as 'a body of thought that challenges the very notion of social science' (Sibeon 2004: 15). Reference here is largely towards the poststructural (Baudrillard, 1983) and the postmodern (Lyotard, 1986). The discussion is not concerned with versions such as those of Jameson (1991) and Harvey (1989) which draw upon Marxism, or others, such as Lyon (1994) and Kumar (1995) that argue 'that we live in a postmodern type of society' (Sibeon: ibid). The focus here is upon the aspects of relativistic, postmodern theory which relate to issues of metatheory, agency-structure, micro, macro, etc; [central concerns of the study] , with the ultimate concern being the development of a metatheoretical framework to study biotechnology.

Sibeon (2004) argues that, despite the claims of Gellner (1993) who sees little of value in postmodern thinking, certain aspects are relevant to the development of theory (Lemert, 1993) and to methodological concerns relating to how empirical studies are designed (Fox, 1991). The author regards poststructuralist and postmodern views of contingency (Bauman, 1992; Lyon, 1994: 4), locale and power to have a part to play in the (re) construction of post-postmodern social theory. Indeed, there are similarities in Bauman's (1992: 192-6) emphasis upon multiple loci of power, contingency, time-space variability, cultural pluralism etc as characteristics of postmodernity, and the anti-reductionist, 'flexible–but–realist social ontology' (Sibeon ibid: 17) of Roger Sibeon's original framework, which provides the basis of the modified framework employed here for the study of biotechnology. Sibeon (ibid) is in favour of 'putatively non-reductionist postmodern notions of social indeterminacy and time-space variability', provided they are modified along the lines of 'realist' social ontology. They are seen as useful and relevant not only to sociology but also to political science (Marsh and Stoker, 1995; Hay, 2002) and to policy analysis (Fox and Miller, 1995; Rhodes, 1997). Sibeon suggests that postmodern theory has certain merits in the direction of attempts to criticise 'modern', essentialist theories and the knowledge-claims of meta-narratives, and indeed theories of any kind which involve reductionism, reification, functional teleology and essentialism. In general, however, he is of the view that postmodern thought cannot provide a cogent and viable theoretical or methodological basis for social analysis.

It is possible to identify the main deficits within postmodern theory, without wishing to embark upon an extended critique, in so far as they relate to the theories of the modified framework employed here, such as essentialism, reification etc. Firstly, when postmodern critics criticise 'modern' social theories, they tend to equate the latter with the problems associated with the four 'cardinal sins' of reification, reductionism, essentialism and functional teleology. However, one of Sibeon's tasks has been to develop forms of reasoning that avoid the four general 'sins'. The modified framework employed in later Chapters to examine selected examples of biotechnology develops Sibeon's framework further to regard *genetic fatalism*, which is 'a pervasive tendency within contemporary social science to equate genetic predisposition with inevitability' (Powell and Owen, 2005: 28), and *the oversocialised gaze*, which refers to 'the 'environmentalist' accounts characterised by a strong antipathy towards genetics or even partially genetic explanatory frameworks' (Powell and Owen, ibid) also as 'cardinal sins' of illegitimate theoretical reasoning. Secondly, as Sibeon (ibid: 18) suggests, there is an 'epistemological contradiction' between the postmodern as a '*type of society*' and the postmodern as a '*type of theory*'. In the first example, Sibeon suggests that postmodern writers are asserting that there is something 'real'in the former mode (empirically), exhibiting, 'definite structural characteristics (consumerism, extreme cultural pluralism and diversity, social fragmentation, individualism, informationalism, and so on)' (ibid), and in the latter mode, they claim that 'society is never something that is real but is, rather, an effect of discourse'. So, we are being told, simultaneously, that on one hand postmodern society 'exists', and on the other that it does not. Although postmodernists welcome contradiction and incoherence, there is a problem for those who seek to develop *meta-theory* here. It revolves around attempts to combine realist ontology with idealist epistemology. Third, as Best and Kellner (1991) suggest, postmodern theory tends to underestimate the significance of interpersonal relations. Thus, it could be argued that postmodern theory neglects subjectivity and also intersubjectivity. As Sibeon (ibid) suggests, postmodern theorists have a tendency to lack an adequate theory of agency and of the actor or agent and subscribe to a flawed conception of agency-structure. This is of some importance, as the author argues that interaction and intersubjectivity mediate the relation of agency to structure. Sibeon goes on to argue that postmodern theorists tend to employ deterministic conceptions of the social actor or agent. He sees these conceptions as 'curiously reminiscent' of the determinism (social) and reification of such reductionist, 'modern' theories as Marxism, radical feminism and Parsonian structural functionalism (ibid). The postmodern claim that knowledge and theories 'embody a standpoint' that has been 'determined by the structural location' of the theorist in terms of his/her membership of social categories is, according to Sibeon (ibid), a 'determinist or reductionist' conception of agency. In this flawed conception of agency, actors are seen as products or effects of discourses, determined by social contexts. Postmodern theory reproduces the 'cardinal sins' of reification, reductionism, essentialism and functional teleology, and entails an 'objectionist conception of agency and structure' that does not fit neatly alongside an 'idealist epistemology' (ibid). Fourthly, Sibeon cites relativism and the postmodern view of the relation of discourses to social contents as contradictory. Here Sibeon highlights *relativism* as a major deficit in postmodern theory, and 'elevates' it to the 'status' of a 'cardinal sin'. In the modified framework employed here, *relativism* is most definitely codified as a 'cardinal sin'. Seidman (1994: 325) represents a tendency within postmodern theory to claim that, 'society or social contexts determine actors' discourses'. This claim is very similar to theories of structural predetermination which posit that actors' cognitive processes are largely determined by their structural locations, and by specific social-historical circumstances. Alternatively, postmodernists claim that 'discourses do not refer to any ontologically prior or pre-discursive social reality', as no 'reality' exists under postmodern conditions; instead discourses 'produce or construct social reality' (Sibeon, ibid: 19). As the author suggests, there appears to be a contradiction here. Postmodernists are arguing both that discourses *produce* or 'determine') social contexts, and that discourses are *determined by* social contexts. Sibeon argues that it is 'appropriate' here to endorse the views of Berger and Luckmann (1971), White (1992: 305) and Law (1986: 304; 1991a: 18), each of whom regard the 'relation of discourses to social contexts' as 'dialectical', though 'without properly recognising that the relation is often only *loosely* dialectical' (Sibeon, ibid). It could be argued that there is no automatic, direct link between action and structure.

As suggested earlier, while certain aspects of postmodern thought contain some useful analytical precepts, especially in relation to contingency and the 'healthy' aspect of incredulity towards metanarratives (Lyotard, 1986), the genre as a whole is regarded as 'seriously flawed' by Sibeon (ibid) and, others such as Rojek and Turner (2000). Sibeon suggests, for example, (whilst acknowledging that there are differences of emphasis among postmodern critics of sociology) that the 'anti-foundationalist' form of postmodern theory (Lyotard, 1986) is 'itself a metanarrative' (ibid). Such relativistic, anti-foundationalism is considered by Turner (1990: 248) to be 'unduly deterministic', failing to analyse the 'epistemological incommensurability between the postmodern as periodisation and as theory'.

Rojek and Turner (2000) outline a critique of 'decorative sociology' as a trend in contemporary social science where 'culture' has eclipsed 'the social'. This is in line with Sibeon's (2004: 22) identification of 'a somewhat whimsical aestheticism that enjoyed its heyday at the height of the postmodern turn' during the late 1980s and early 1990s. Rojek and Turner (ibid: 1) use the term 'decorative sociology' to refer to, 'a branch of modernist aesthetics which is devoted to a politicised, textual reading of society and culture'. The authors seek to 'challenge the political self-image of decorate sociology' as a contribution to political intervention. Rojek and Turner acknowledge the contribution of 'the cultural turn' to a revision of the relationships between identity and power, race and class, ideology and representation etc. However, the authors emphasise that it has done so 'chiefly at an aesthetic level'. Influenced by Davies (1993), they argue that the greatest achievement of the 'cultural turn' has been to teach students how to read material politically. Their hypothesis is that what might be called the 'aestheticisation of life' has not translated fully into the politicisation of culture. Similarly to the calls for a return to sociology theory and method (Sibeon, 1996, 1999, 2004, 2007) and Mouzelis (1991, 1995), Rojek and Turner argue that an 'adequate cultural sociology' would have to be 'driven by an empirical research agenda', embrace historical/comparative frameworks and focus upon 'the changing balance of power in Western capitalism' (ibid). Rojek and Turner maintain that there is a case for advocating the importance of theoretical cumulation as opposed to anti-foundationalism. The authors equate cultural relativism with moral relativism. As they suggest (ibid: 645)....

Because postmodern cultural studies assumes moral relativism, it cannot produce, let alone accept, a unified moral criticism of modern societies. As a result it is intellectually unlikely that cultural studies would develop an equivalent to Weber's notion of rationalisation or Marx's concept of alienation. Postmodern cultural studies finds it difficult to promote a political version of the modern world apart from an implicit injunction to enjoy diversity and resist commercialisation of popular culture.

It is necessary here to examine postmodern theorists' portrayal of their targets. Postmodernists have tended to depict sociology as dogmatic and over-confident in their rejection of sociological reliance upon metatheoretical, theoretical and methodological foundations. Gurnah and Scott (1992: 41) for example, argue that 'the whole of post-war European and American sociology' exhibits a 'degree of superficiality and smug sense of infallibility'. In response Sibeon (2004: 22) argues that sociology in its modern form is actually characterised by 'theoretical pluralism and by epistemological uncertainty and reflexivity', and that these characteristics have accelerated since the 1950s. The work of Gibbs (1989) also supports this claim. According to Swingewood (2000), the history of sociological thought is characterised as much by intellectual caution and introspection as it is by smugness and certitude (Calhoun et al. 2002). As Sibeon points out, Seidman's (1992) argument along the lines that to formulate sociological perspectives based upon explicitly articulated theoretical or metatheoretical premises ('foundations') is the same as suggesting that all 'competing sociological perspectives' should be denied the right to exist (Sibeon: ibid). Seidman goes on to argue that 'foundational rationales are never more than local, ethnocentric prejudices' (1992: 60). As Sibeon (ibid) points out, Seidman then proceeds to develop a 'foundational rationale' for his own views.

It is arguably possible to avoid the four 'cardinal' sins referred to by Sibeon, without having to engage in postmodern hypercontextualism. In later Chapters we see the development of a flexible, meta-theoretical framework employed to examine aspects of biotechnology, and the desire here is to follow Sibeon in avoiding the crude either-or choices between ontologically rigid and dogmatic 'modern' theories, and the equally flawed relativistic approaches of postmodern and poststructuralist theories. The approach advocated here follows Sibeon's work and entails a pliable ontology which includes a processual conception of 'the social' as potentially variable across time and social space. The ontology, though flexible, is 'realist'. That is to say that it favours a conception of a heterogeneous, shifting empirical reality 'out there', which rejects the hypercontextualism and relativism of postmodern/poststructuralist epistemology. As referred to earlier, Lyotard's postmodern repudiation of 'metanarratives' is, in part, a rejection of reductionist, single-order 'modernist' theories which emphasise, for example, capitalism or globalisation as universal principles of explanation. It may be possible to agree with Lyotard's criticisms of 'modern', essentialist theories and to borrow his sense of 'conceptual flexibility', whilst at the same time insisting upon approaches based upon ontological and epistemological 'realism', which are rejected by postmodernists. In Sibeon's (ibid:23] view, therefore, 'a pliant but realist ontology' would recognise that we can conceptualise foundations in terms of (meta) theoretical postulates pertaining to social processes and mechanisms to do with agency, structure, time-space and social chance rather than to empirical structures, events or social patterns. Sibeon considers Giddensian structuration theory to be 'ontologically flexible' (ibid), referring to Cohen's work (1987: 279-80, 285, 289). Cohen (1987: 297) observes that, in regard to Giddens's assumptions about macro social processes: 'consistent with the ontological flexibility of structuration theory at large, Giddens holds open for substantive enquiry all questions regarding specific systemic patterns' as well as the degree to which social systems are, 'stable, organised and permeable'. The type of ontological 'relation' which Cohen refers to here, as does Stones (1996), is not the same thing as what Sibeon (ibid: 24), calls, 'crude modernist theories' which replicate the 'cardinal sins' of reification, reductionism, essentialism and functional teleology.

Postmodern criticisms of foundations in sociological theory relate to questions raised concerning the relation of social discourse to social contexts. Sociological theory is arguably a discipline and subject in its own right, and as such may be modified and improved upon. As Sibeon (ibid: 4] makes clear, this does not have to conflict with the 'parallel view that sociology does not stand in grand isolation' from 'lay' discourses (Friedrichs, 1972: 298). It can be argued that the relation of sociological discourse to 'lay' discourse and to social conditions is 'loosely dialectical' (ibid), described by Giddens (1987: 32) as 'the double hermeneutic'. There is a suggestion here, by Sibeon, that the concepts/theories of sociology 'spiral' in and out of social life. As the author suggests, when sociologists investigate the 'empirical world', actors' meanings and concepts 'will tend to wind their way' into sociological discourse. Sibeon (ibid), cites juvenile crime as an example, explaining how studies of the sociological area draw upon/refer to concepts of criminality, self-responsibility etc that are employed by 'relevant actors' such as young offenders, police officers etc. Actors' meanings, therefore, 'permeate, and are crucial resources for, social scientific accounts' of the empirical world (ibid). Sometimes, sociological concepts are 'diffused' to settings outside of academe. Mouzelis (1995: 52) has noted, according to Sibeon (ibid), the extent to which particular social science discourses have/do not have impacts upon social contexts, as 'an empirical variable', though this is something postmodernists and poststructuralists might question in view of their belief that the social world described by academics is entirely a product of discourse. Sibeon goes on to say that, while there may be no 'direct link' between academic and 'lay' discourses, there 'tends to be a *loosely dialectical* linkage' to the extent that each kind of discourse influences the other to some degree, which is 'a matter for empirical assessment' in each case (ibid: 25). Here, we must note that Sibeon refers to a theory which forms part of the discussion of agency and social actors later in this Chapter; the idea that the relation of social science discourse to social contexts is such that each is partly shaped by the other but neither fully determines the other; each has (relative) autonomy. An acknowledgement of the idea that social science discourse 'spirals' out and into the 'social fabric' is *not* to suggest that social scientists cannot formulate general propositions about social life (White, 1992: 5, 304). Rather, Sibeon argues that we should reject the relativism of Gurnah and Scott (1992) and Seidman (1992; 1994), reflecting that, the extent to which such propositions embody (or do not, as the case may be) any identifiable cultural, regional, ethnic, political or gender-related nuances or specificities of time and place, and, where any of these are shown to exist, their implications for social science knowledge, are matters that in each instance merit collaborative academic assessment and reflexive response. Taken together, these considerations suggest that archetypal absolutist and positivist scientism, or the hypercontextualism and relativism of poststructualism and postmodernism, are not the only epistemological alternatives: relatively impersonal generalised categories of social science knowledge are possible (Alexander, 1992: 323) (Sibeon, ibid: 25)

In relation to the controversies surrounding the concept of theoretical 'foundations', a very important proposition (in epistemological and ontological terms) is that meta-theory, substantive theories and empirical data should be compatible with each other, and should influence each other. Indeed, it could be argued that 'meta' or sensitising theory, such as Sibeon's original anti-reductionism, or the modified version applied here to study biotechnology, should be seen as *revisable*, 'at the point where substantive theory and empirical data intersect', and 'the relation of this intersection to metatheory should be regarded as a dialectical relation' (ibid). In theory, then, 'theory-data reciprocity' should stand in a, 'mutually regulating relation to metatheory', and theory, data and metatheory should 'shape each other' (ibid). In certain respects, Sibeon's original anti-reductionist meta-theory, from which the framework employed here develops, is similar to Margaret Archer's (1995). Archer's 'morphogenetic' social theories are discussed in greater depth later in this Chapter as relevant to the transitions in social theory that have led to post-postmodern 'synthetic' approaches. Archer, as Sibeon (ibid: 26) makes clear, makes a distinction between (her terms) Social Ontology, Explanatory Methodology and Practical Social Theory. Firstly, she suggests that these should be 'mutually regulative' and consistent (Archer 1995: 28), and that it is an advantage to have a clear social ontology/metatheory as metatheoretical assumptions of the ontological kind are impossible to avoid and influence substantive theory as the 'likelihood of ontological contradictions' (Sibeon, ibid: 26) arising at the level of substantive theory is reduced by an explicitness at the level of metatheory. As Archer has observed, (which Sibeon (ibid) concurs with), the possession of an 'explicit' epistemology does not guarantee against all errors. Archer warns against 'conflationary' conceptual schemes; there may be a consistency there between social ontology, methodology and 'practical' social theory' yet because they are 'conflationary', such schemes can be of no explanatory use. 'Conflation' refers to the compression of two or more variables, so that their individual idiosyncratic qualities are obscured. Sibeon's original anti-reductionist framework, focusing upon agency-structure, micro-macro and time-space, and are arising out of a critique of reductionism, essentialism, reification and functional teleology has an 'affinity' with Archer's framework, and also with Layder's (1998a, 1998b: 92) theory of social domains. Archer's 'analytical dualism', opposed to the duality of structure and action, is similar to Sibeon's position regarding 'duality of structure' in that it seeks to maintain the distinction between agency and structure. This 'duality of structure' concept is included in the framework developed in later Chapters to examine biotechnology.

This Chapter deals with some of the core concepts and concerns of Sibeon's original framework. Sibeon's original framework incorporates Foucauldian insights pertaining to power. Whilst it could be argued that the main focus is upon agency-structure, micro-macro and time-space, Sibeon (1996, 1999, 2004) certainly draws upon Foucauldian ideas pertaining to power. However, whilst acknowledging that Foucauldian (1972, 1980a, 1982) concepts relating to power, if used selectively, can serve as a useful corrective against exclusively systematic perspectives (Sibeon 2004: 135), there is a tendency in Foucault's work (1972, 1980a, 1980b, 1982, 1991) towards *relativism*, and in particular, a 'failure to incorporate a non-reified concept of agency into social analysis as part of a dual system and integration approach' resulting in illegitimate teleological forms of explanation (Sibeon 2004: 73). The point of this section of the Chapter, concerning meta-theory and problems associated with the 'cultural turn' is to highlight problems around *relativism* in particular, and to emphasise the need to move beyond postmodern and poststructuralist relativism towards a *post-postmodern* framework. Drawing upon Sibeon (2004), Owen (2006a: 193) suggests that 'Foucault's conception of *power* may be successfully incorporated' in analysis and meta-theory, and this must be done in a critical and selective fashion....

> informed by the critique of *agency-structure, micro-macro* and other terms offered in the 'new' framework. If we are to employ Foucauldian insights pertaining to *power* we must recognise the *dialectical* relationship between *agentic* and *systemic* forms of power; the *relational, contingent* and *emergent* dimensions of power, and the concept, that *contra* Foucault, aspects of power can be 'stored' in positions/roles and in social systems/networks (Mouzelis, 1995). (Owen, ibid).

It is necessary here to closely examine Foucauldian analysis, both because it employs a *relativism* which the framework here, and the frameworks employed in the work of Owen (2006a, 2006b) reject, and because, as stated previously, Foucauldian notions pertaining to *power* may be employed in a modified form to study selected examples of biotechnology. However, here we must begin to consider the usefulness of Foucauldian insights in relation to the development of the framework, and to a study of the Genome Project itself.

THE HUMAN GENOME PROJECT AND FOUCAULDIAN 'POWER'

It could be argued that we can view the Project, in terms of *power*, 'as an example of Sibeon's (1996: 52) 'emergent' power (Owen, 2006b: 909). If we analyse the Project, drawing upon Callon and Latour's (1981) 'Actor-network' theories and Foucauldian socio-political perspectives, we might conclude that power is not a fixed capacity that, 'inheres in an actor by virtue of the actor's structural location within an overarching social system' (Sibeon: ibid). Rather, we might consider that power is at least partly *emergent*, in the sense of being an *outcome* of social interactions. As Law (1986: 5) suggests, power is in part an 'effect', not a cause of strategic success achieved by actors during the course of their interactions with other actors in specific situations. When examined in relation to the flexible framework outlined in the introduction, and at the beginning of this Chapter, it could be argued that 'Foucauldian insights (1972, 1980, 1982) contain an anti-foundational *relativism* that should be repudiated' (Owen, ibid).

Foucault's 'overly-holistic' concepts (Sibeon 2004: 71) have strongly *essentialist* overtones, which fail to address and acknowledge the discontinuities between macro-cognitive and micro-behavourial spheres of the social world (Sibeon ibid). There appears to be a tendency in Foucauldian theory to 'collapse' the distinctions between *agency* and *structure* and between *micro* and *macro*, resulting in what Archer (1995) calls 'central conflation'. Foucauldian insights tend to rest upon a '*reified* concept of *agency*' (Sibeon, ibid: 74). Foucault's theorising tends to incorporate 'reified concepts of agency and the concept of *duality of structure* into analysis, with the result of creating *teleological* forms of explanation (Owen, ibid). Armstrong (1987: 1217), drawing upon Foucauldian insights, warns against too readily endorsing alliances between 'bio-social perspectives', perhaps displaying the *genetic fatalism* and *over-socialised* approach of such Foucauldian analysis' (Owen ibid). Foucault's (1972) suggestion that sexuality is purely a socio-cultural creation is explored in greater detail in Chapter Three, with contrary evidence provided by the work of Hamer and Copeland (1999). However, despite some of the theoretical deficits outlined, *power* is one aspect of Foucauldian analysis, which, 'if used in a 'modified and reflective way, can certainly contribute towards the sociological study of biotechnology' (Owen, ibid: 910). *Power* in the Foucauldian sense, is, 'the milieu in which individuals and groups operate; the workings of power are not centred in any one group or source, and do not arise from any given location in the social structure or operate from any singular site' (Owen, ibid). Power is e*verywhere* – the network of surveillance, patterns of discipline and knowledge that serve them have emerged outside of any social actors' control. These ideas can be used as a corrective against exclusively systemic perspectives.

One concurs with Sibeon (2004: 135) to the ends that the Foucault-influenced, 'Actor-Network' theories of Callon and Latour (1981) contain persuasive arguments in favour of power having, 'no single or prime cause, but that strategic success in the acquisition of power is always potentially reversible'. For Sibeon (ibid), we should borrow elements of Foucault's ideas pertaining to power in a, 'critical, selective fashion' informed by critique of *agency-structure, micro-macro* and *time-space*. It is useful to 'recognise the relational and emergent aspects of power, but power also has *systemic* qualities' (Owen, ibid: 23-24). Latour, influenced by Foucauldian insights, claims that, 'power is not something you may possess or hoard,' and that power is 'an effect....never a cause' (1986: 265). Sibeon (ibid: 136) cogently disagrees with Latour's suggestions, arguing that, 'Latour is wrong; power can be hoarded or stored, and therefore power – though often an effect – can sometimes be a 'cause''. One tends to agree with Sibeon that Foucault and 'Actor-Network' theorists such as Callon and Latour (1981) and Latour (1986) tend to, 'push their relational and processual conception of power to the point of denying that power can be 'stored' in roles and in social systems and networks of social relations'. As will be seen in Chapter Three, we may therefore incorporate a synthesis into the meta-theoretical framework which combines Foucauldian and other *relational* concepts of power with an understanding that there is also a *systemic* dimension to power. An acknowledgement that 'some social agents 'possess' more power than others (Best and Kellner, 1991: 70), and that the reason for this may lie in the fact that certain elements of power can be 'stored' in roles, social systems etc, is something which Foucauldian theory tends to ignore' (Owen; ibid].

Particularly useful for the sensitising framework developed here is the idea that there may be multiple forms of power including systemic power (associated with power storage) in discourses, social institutions, social positions/roles, and in social systems), and agentic power which refers to a capacity of agents. These *systemic* and *agentic* powers should be conceptualised as autonomous, though they may influence each other and agentic power....

> may be of a relatively contingent, emergent form. Such agentic power may interact with systemic power (role/position) that is 'stored' within what Foucault terms 'discourse', and social systems. In other words, we may think in terms of a *dialectical* relationship between *systemic* and *relational* forms of power (Owen, ibid: 910-911].

Regarding an analysis of *power* in relation to the 'public' Human Genome Project, we must examine certain aspects of how such power is created and maintained. We need to examine the 'genealogy' of existing relations, how they have emerged and the discourses they reflect/reinforce. The work of Owen (ibid) has made a cautious attempt to do so, and it has been argued that social actors have two related qualities. They have an ontological status as actors (that is, in principle they are able to formulate and take decisions and act upon those decisions), but secondly, this may be an *intermittent* status that does not obtain in all circumstances. However, in the case of the Human Genome Project, it can be argued that it is an entity that is *not* an actor. To suggest otherwise would be to engage in *reification*. As Owen (ibid: 911) has suggested....

> The government of a group of 'nation states' may decide that it is desirable to create a project such as the Human Genome Project that is 'empowered to take decisions and act on the collective behalf of its member states'. (Sibeon 1996: 57). In this case, the Human Genome Project may be regarded as a *supra-national organisation*. We must keep firmly in mind a crucial component of the framework's methodology - Hindess's (1986: 115) formulation of social action, in which he defines an *actor* as a 'locus of decision and action where the action is in some sense a consequence of the actor's decisions'.

If we take Bennington and Harvey's (1994) idea of transnational and inter-regional policy networks composed of governmental and non-governmental actors and 'apply' it to the 'public' Human Genome Project as a whole, 'we can identify the International Human Genome Sequencing Consortium (HGSC) as a *policy network'* (Owen: ibid). It is composed of more than 2000 scientists at 20 institutions in six countries (Dennis and Gallagher, 2001), and functions to develop the technology required to characterise the genomes of humans and other organisms, and to address the ethical, legal and social implications of this 'new' information. The institutions range from the governmental (US Department of Energy) to the non-governmental (RIKEN Genomic Sciences, Japan).

We need to acknowledge the multiplicity of forms that *power* takes, in relation to the Human Genome Project. This includes *systemic* power 'stored', '*contra* Callon and Latour (1981), in discourses pertaining to the Project, the twenty or so institutions that make up the Human Genome Sequencing Consortium (HGSC), the social positions/roles of actors and *agentic* power which refers to a capacity of actors' (Owen, ibid: 26-27).

Hopefully, at this point, the need to move 'forward', beyond the relativism and nihilism of the 'cultural turn', and the deficits of Foucauldian analysis identified here, towards a meta-theoretical framework which focuses upon *agency-structure, micro-macro,* aspects of Foucauldian *power* and *time-space* has been established. In Chapter Three, the argument for including a focus upon *biological variables* (evidence from behavioural science and evolutionary psychology which supports, at least in part, biological causality regarding some human behaviour is explored. In what follows, we examine in greater depth *agency* and *structure* and *micro* and *macro,* together with the links between the theoretical approaches of certain contemporary social theorists (Roger Sibeon, Derek Layder, Nicos Mouzelis and Margaret Archer) and the implications for both post-postmodern sociology and for the development of the framework. Chapter Two contains a codification of a suggested metatheoretical framework, which is 'applied' to selected aspects of human biotechnology in the final (Five) Chapter.

Agency and Social Actors

In what follows, the intention is to establish in general terms the importance of the concepts of *agency-structure* and *micro-macro* in terms of how they refer to differing dimensions of social reality, and to review certain conceptual dilemmas and arguments surrounding the contemporary debate of these concepts in social science. On the basis of a critique, suggestions are offered as to how we might employ these concepts in a developed framework. We have already cautiously criticised the relativism of the 'cultural turn' and the Foucauldian concept of *power* previously. It should be noted that we are dealing here with *modern* social theory, and despite the fact that some of the theories draw upon classical theorists (Marx, Weber, Durkheim etc), the intention is to focus upon how *agency-structure* and *micro-macro* are handled by contemporary theorists. Here, we are concerned with major contemporary theorists such as Anthony Giddens, Margaret Archer, Nicos Mouzelis and Derek Layder.

It is arguably not enough simply to say that Lyotard (1986) and other postmodern/poststructuralist critics neglect agency if one is concerned with the development of post-postmodern, metatheoretical framework. As Sibeon (1997a: 5) suggests, what is required is a 'non reductionist and non-reified definition of agency'. In Sibeon's original, *anti-reductionist* terms, agency is defined by means of Hindess's (ibid) definition (discussed previously) in which actor/agent is 'a locus of decision and action', where the action is in some sense a consequence of the actor's ability to formulate and act upon decisions'.

However, as we have previously seen in the case of the Human Genome Project, there are some entities that cannot under any circumstances be or become actors. These include taxonomic collectivities such as 'men', 'women', 'white people', 'black people', etc; social classes; 'society' and 'the state'; and objects such as money, written materials, natural disasters, diseases and so on. Additionally, none of these categories of phenomena can be or become agents.

In social theory and in social sciences disciplines in general, *agency* tends, as Sibeon (ibid) points out, to be associated with 'human creativity' and social chance (contingency), and *structure* with 'patterned relations, with constraints upon behaviour and with macro-social phenomena'. '*Micro*' is 'for the most part employed as a term that denotes settings of face-to-face interaction', while '*macro*' is 'frequently used to refer to 'society' and social institutions' (ibid). Often we may find that *agency* is linked to *micro*, with *structure* associated generally with macro-social phenomena. Sibeon's original anti-reductionist framework views some of the conceptualisations as problematic. Nevertheless, we must recognise that they enjoy quite wide support and there is a general (though not universal) tendency among social theorists to accord considerable importance to distinctions between agency and structure, and micro and macro. As Sibeon suggests, it is the view of certain contemporary theorists that agency-structure (Archer, 1988: ix-x) and micro-macro (Knorr-Cetina, 1981: 2; Munch and Smelser, 1987; Ritzer, 1990) are '*the* core underlying problematics of social theory and sociology' (ibid: 36). The author has no major disagreement with such views, but adds that *'social choice*' (*contingency*) and '*time-space'* are also highly significant concepts. The latter term is incorporated into the framework in Chapter Three, alongside agency-structure and micro-macro. *Contingency* is arguably another term for *social chance*. Sibeon's original *anti-reductionism* portrays social life as relatively fluid, processual and *potentially* indeterminate unless and until segments of social life are spatially/temporally stabilised and institutionalised for short or long durations of time (Hindess 1986: 120-21). This raises an important question. As Sibeon (1991: 68) maintains, there are arguably no valid reasons for constructing an 'either/or theoretical dichotomy between, on the one hand, a structural conception of rigid system determination and predictability' and, on the other, a 'conception of the social fabric as a process of endless flux', and random change. Elias's (1978: 162) figurational sociology paints a picture of social reality as indeterminate, continuous and processional, as an unfolding of 'figurations in constant flux with neither beginning nor end'. One could argue that Elias is utilising an illegitimate dichotomy which is neither statis nor flux. Sibeon's original anti-reductionist framework transcends this dichotomy through the empirical application of a theory which recognises that social life is always potentially open to change and variability, the extent to which stability and continuity, or else discontinuity and change, actually occurs being treated as a variable for empirical investigation. Social conditions are viewed as being *contingently* produced and reproduced as an effect of the self-reproducing tendency of social systems and networks; and in part as the consequences of actors' decisions and actions. As Clegg (1989: 17) indicates, because the social fabric is not structurally predetermined it is, at least, potentially indeterminate and variable. In the concept of '*time-space'* a core assumption is that material which travels in time must also travel spatially; and that spatial dissemination necessarily has a temporal dimension. Cohen (1989: 77) analyses Giddens's structuration theory observing: 'if social patterns are embedded in the reality of social activity, then a concern for time and space becomes difficult to avoid. Social conduct, after all, is always situated in specific settings, and it takes time to engage even in the most fleeting practices, let alone sustained sequences and series of interaction'. Arguably, Cohen's observation highlights the fact that temporality is a very significant variable among competing sociological theories. As Clegg (ibid: 212) suggests, 'Different theoretical perspectives diverge on the temporality within which the conceptualisation of action is achieved'.

As mentioned previously, there are theoretical approaches in modern social theory (some of which draw upon classical sociology) which tend to emphasise *structure* whilst neglecting *agency*, according analytical primacy to the macro-social. *Structuralism* is one such approach. Here, there is an analytical movement from subject/actor to structure. In other words, people are concepualised as 'products' of structure, as 'decentred' from their own meanings. As Sibeon indicates, structuralists view structure rather than agency as the 'main focus of analysis' (2004: 39). The theoretical shift from agency to structure is based on the idea that social structure is self-regulating and synchronic. As Scholes (1974: 10) has said, 'At the heart of the idea of structuralism is the idea of system: a complete, self-regulating entity that adapts to new conditions by transforming its features whilst retaining its systemic structure'. Such a statement entails *reification*; one of the 'cardinal sins' repudiated by Sibeon's original framework, and incorporated into the new framework (codified in Chapter Two). As Sibeon (ibid: 39) goes on to say, this *reification*, together with 'determinism based on a reductionist notion of language and action' is a feature of a subdivision of structuralism, known as structuralist linguistics (Saussure, 1974; Levi-Strauss, 1963; 1974). Sibeon cites the work of Sharp (1980) as an example of structural linguistics' objectivist, reified conception of language. The 'determinate supposition that language is an autonomous entity', existing independently is part of the structuralist idea (found in Barthes, 1967) that human actors are 'inserted' into language (Sibeon, ibid). Similarly, in structural linguistics, there is no 'correspondence between a word and the object to which the word refers' (ibid). Rather, the 'meaning' of a word is determined by its relation to other words, and to concepts of objects, but not to material objects in 'the real world'.

It is not only structuralism and its offshoots which engage in the deterministic neglect of agency. *Poststructuralism* (we have previously discussed Foucault's poststructuralist ideas pertaining to power) is also implicated. The work of Lacan (1977) and Foucault (1970; 1972; 1980a) are offered as examples by Sibeon. Structuralists argue that, 'neither signifier nor signified represent or are linked to anything 'real', but second, that there is a more or less stable link between signifier and signified' (Sibeon ibid: 39). Poststructuralists, such as Lacan and Foucault would reject the second proposition. According to poststructuralism, texts and indeed human society are 'open' to a multiplicity of 'readings', and 'any interpretation or reading is never 'finished' or 'final' but is and should be subject to endless challenges and re-interpretations' as can be seen in the work of Derrida (1982) (ibid). 'Reality' itself is denied, and analysis must centre upon discourses and the construction of a 'putative reality' (ibid). Sibeon gives the example of the poststructuralist concept of *intertextuality*, which refers to the 'interplay of texts, which results in texts modifying each other when they are read and re-read (re-interpreted)' and thus texts, and by implication any 'discourse' that refers to the 'real' world, 'are indeterminate and subject to continuous re-interpretation and reformulation', and nor is there a 'correspondence theory' criterion for arbitrating among interpretations or discourses' as poststructuralists reject the idea that there is a 'real world' against which 'competing discourses or representations can be empirically tested' (ibid: 39-40). As noted in the sections of this Chapter concerning *'Metatheory and Problems of Relativism'* and *'The*

Human Genome Project and *Foucauldian Power*, postmodern and poststructuralist critics not only engage in *relativism*, but their approach to *agency* is, arguably, deterministic, viewing the social actor as 'decentred', as a 'product' of discourse. Sibeon (ibid: 40] gives Laclau and Mouffe's (1985; 1987) work as an example of *Discourse Theory*, which borrows in part from Saussure's structuralist linguistics. This latter form of theory, drawing upon postmodern and poststructuralist insights posits that discourses and societies are never 'closed' – 'there are no 'fixed' or stable meanings – but rather, they exist in a state of endless reformulation and flux (Sibeon, ibid: 40). *Discourse Theory* includes the following poststructuralist ideas: 'there is no pre-discursive social reality and therefore no benchmark against which we can empirically test the veracity of competing discourses or representations of 'reality'' and 'any discourses which attempt to convey a sense of the 'real world' are arbitrary, and endlessly reformulable' (ibid). Actors' forms of thought and actions are merely the 'products' of discourse. One tends to agree with Sibeon's criticism of such concepts, not only in terms of the neglect of agency and social determinism on the part of *Discourse Theory*, but also because of its refusal to take seriously the concept that there exists an ontologically complex and real empirical world that is at least partly independent of discursive activities.

There are other examples of contemporary theorists (deterministic approaches) that tend to either neglect agency, or elevate concepts of structure above agency. Sibeon (ibid: 40-41) describes Habermas's (1986; 1987; 1989) theoretical schemes as another example of an 'over-emphasis on macro-structural phenomena and a neglect of agency'. According to Sibeon, Habermas attempts to 'combine voluntaristic action theory with a deterministic systems theory'. Whilst acknowledging that 'non-contradictory syntheses of action and structure are possible', Sibeon (ibid) claims that Habermas does not achieve this. Because Habermas's (1987: 159-60) 'systems are imbued with agency', they are thereby *reified* (ibid). Further, the author draws attention to....

> Habermas's contradictory tendency to largely ignore agency in the study of what he calls 'system' (polity and economy). In his theoretical framework, system is inspected only in system-integration terms, whereas he argues, the 'lifeworld' (taken for granted meanings employed in everyday life among family, friends and associates) should be viewed in social-integration terms: this neglect by Habermas of agency and actor-actor relations in politics, bureaucracies and markets and conversely, the neglect of systemic, role-patterned, and institutional aspects of the lifeworld, limits the usefulness of his approach to agency-structure (ibid: 41).

Viewed in terms of the definitions of *agency-structure* and *micro-macro* used by Sibeon, and incorporated into the 'new' modified framework employed here, Bourdieu (1977; 1984; 1990; 1998), widely regarded in his lifetime as a major theorist of agency-structure, is 'ultimately condemned to a deterministic view of agency' (Sibeon, ibid). Bourdieu's theory of 'constructivist structuralism' is 'much closer to structuralism than to constructivism', as he believes 'actors' dispositions are largely determined by the social positions they occupy and by their 'habitus' (ibid).

In what follows, we address some themes, controversies and links between the approaches of a selection of contemporary theorists pertaining to debates around *agency-structure* and *micro-macro*. As Sibeon (2004: 45) has observed, 'a certain amount of ambiguity and also variation in meaning are features of the debate surrounding agency-structure and micro-macro', and 'misinterpretations occur quite often'. One example is Mouzelis's (1991: 35) criticism that Giddensian structuration theory ignores 'macro-action' (which Sibeon (ibid) describes as meaning 'decisions made by people in positions of authority such as government ministers or heads of large firms' as well as 'decisions made by meetings or committees whose members have a high level of formal authority' such as meetings of heads of state). Sibeon (ibid) argues that Mouzelis's criticism, repeated in his (1993a: 682) later work, rests on a misunderstanding on the part of the critic himself. According to Sibeon (ibid), Mouzelis's critique is flawed due to his misunderstanding of 'how the theory of structuration relates to what Mouzelis (but not Giddens) terms the 'micro-macro' distinction: nowhere, despite Mouzelis's claim to the contrary, does Giddens deny the existence of what Mouzelis rather confusingly calls 'macro action''. Another example of misinterpretation in contemporary debates around agency and structure lies in what Sibeon (ibid) refers to as 'McLennan's (1995: 121) erroneous claim - made in a paper that in other respects is scholarly and insightful - that Mouzelis (1991) criticises Giddens on the grounds that Giddens associates micro with voluntarism and with agency, and macro with structure viewed as constraint upon agency'. As Sibeon makes clear, Giddens (1984: 139; 1993: 7) rejects the conflation of micro-agency and macro-structure, and 'it is also the case that Mouzelis (1991: 32) recognises that Giddens opposes these conflations', and here McLennan seems to have misinterpreted the theoretical positions of both Giddens and Mouzelis.

Aside from differing ways of dealing with *agency-structure* and *micro-macro*, and misapprehension of terminologies, there are 'divergences' that rest on 'real differences of approach at the level of social ontology' in the case of a selection of contemporary theorists examined here (Sibeon 2004: 46). Sibeon identifies the morphogenetic social theory of Margaret Archer as containing an 'inadequate conception of micro-macro' (ibid). The majority of contemporary theorists appear to use 'micro' to refer to small-scale social phenomena or units of analysis. In contrast, Archer (1995: 8-8, 12) argues that micro-macro, 'should not refer to differences in the absolute 'size' of social phenomena, but rather, to *relative* differences in size and to a relational conception of scale associated with the concept of 'emergence'' Sibeon (ibid). Archer (ibid) is suggesting that a given unit of analysis may be 'micro' in relation to one aspect or stratum of society, and 'macro' in relation to others: 'what justifies the differentiation of strata and thus use of the terms ''micro'' and ''macro'' to characterise their relationship is the existence of emergent properties pertaining to the latter but not to the former' (Archer 1995: 9). Thus, as Sibeon (ibid) points out, in Archer's terms, a dyad can be viewed as micro, but if it formed part of a 'slightly larger social grouping (a committee or a household, say), then the latter, in relation to the dyad, would be investigated as a 'macro' phenomena'. Sibeon recommends resisting Archer's conceptualisation of *micro-macro* (though he has 'no quarrel with the general idea of emergence') in favour of more conventional definitions. A reason for Archer's desire to abandon such conventional distinctions between micro and macro lies in her belief that *'micro-macro'* and *'agency-structure'* are 'simply versions of the same debate' (1995: 7). Sibeon (2004: 47) disagrees, arguing that 'micro is not at all the same thing as agency, and macro is not the same thing as structure'. If we draw upon critique of the four 'cardinal sins' of the author's original framework, employing a flexible social ontology, we clearly see that….

> agency is defined in terms of an explicitly anti-reificationist and minimal concept of actor; social structure is defined, in similarly 'minimal' fashion, as 'social conditions' (or the

> 'conditions-of-action'); and micro-macro refer to the units and scale of analyses concerned with the investigation of varying temporal and spatial extensions of the social. In particular, it was emphasised that social structure is a contingently reproduced set of social conditions, not a necessary effect of the social totality or something that is historically predetermined (ibid: 57)

As Sibeon (ibid: 47) suggests, another motive behind Archer's unorthodox handling of *micro-macro* is her assumption (1995: 10-11) that those who use the term 'micro' to refer to small-scale interaction and phenomena, 'tend to regard the micro-social sphere of interpersonal relations as insulated from the macro-social sphere'. Sibeon cites Knorr-Cetina and Cicourel (1981); Munch and Smelser (1987); and Ritzer (1992) as examples of theorists who employ the micro-macro distinction to indicate differences in properties/scale of phenomena, whilst insisting on 'precisely the idea of links between micro and macro' (ibid). Most theorists appear to be opposed, therefore, to what Archer (1995: 10) refers to as an '"isolated" micro world'. Whilst acknowledging the contribution Archer has made, especially with regard to her morphogenetic social theory, Sibeon recommends that we reject Archer's version of micro-macro. The *'micro-macro distinction'*, provided it is not employed in the same way as in Archer's work, and in ways which avoid the 'cardinal sins' of *reductionism, reification, essentialism* and *functional teleology* is....

> a useful conceptual tool for marking out variation in the properties and temporal and spatial scale of social phenomena, for denoting corresponding differences in the nature and size of the units of analysis employed in social enquiry, and the exploring links between these units of analysis (ibid: 48).

Archer (1988, 1995, 1998) is very much involved in contemporary debates around 'dualism versus duality', which, as Sibeon (ibid) indicates, 'centres mainly though not exclusively upon critique of Giddens's theory of structuration (Giddens, 1981; 1982; 1984; 1991b; 1993)'. In what follows, we examine the contribution to this debate, which is in itself related to agency-structure and micro-macro, by selected contemporary theorists. In Chapter Three, the term *'duality of structure'* is 'officially' included in the 'new' modified version of Sibeon's original framework. Certainly, Sibeon most definitely favours *dualism* rather than *duality* in relation to *agency-structure* and *micro-macro.* In the following Chapter, there is a more in-depth consideration of *dualism* and *duality,* the implications for the framework, and an application of the concept of *dualism* to the 'building of bridges' between sociology and biology. It will be remembered, from the introduction and the beginning of the Chapter, that the intention here is to develop Sibeon's framework further so that it entails focusing (amongst other things) upon the *biological variable* (evidence for, at least in part, biological causality regarding some human behaviour) alongside *agency-structure, micro-macro,* and *time-space.* In Chapter Three, we identify a tendency to argue for a *duality of structure* amongst writers of the 'embodied' school, in which 'biology' and 'society' are seen as 'two sides of the same coin'. For example, Shilling's (1993) analysis of the body as *simultaneously biological and social.* Newton (2003: 35) sees 'no epistemological reason to erect a barrier between the biological and the social'. These positions, in the field of those who seek to build links between the social and life sciences explored in the next Chapter, appear to be very similar to Giddens's (1981; 1982; 1984; 1991b; 1993) theory of structuration in the sense that they are 'elisionist' (Archer 1995; 1993), collapsing 'biology' and 'society' together as Giddensian 'flat' social ontology collapses the micro-macro distinction. Giddens is not the only theorist to effectively collapse this theoretical distinction. As Sibeon (2004: 48) makes clear, so do others such as Elias, Foucault, Callon and Latour, and poststructuralist and postmodern theorists. Giddens has suggested that 'tensions' between micro-theorists and macro-theorists arise out of a 'phony war' (1984: 139) and that micro-macro is a false and unhelpful dualism (1993: 4) 'that polarises social scientists into proponents of two opposing approaches' (Sibeon, ibid); he also thinks that the distinction helps perpetuate a tendency of some theorists to link micro with agency, and macro with the constraints 'applied' by social structure upon agency (1984: 139; 1993: 7). Another factor in Giddens's unwillingness to employ these terms is that 'micro-macro' distinctions are dualisms serving to emphasise differences between small groups and larger phenomena, such as organizations. Archer's work is similar, but 'she rejects Giddens's elision and substitutes her 'idiosyncratic version of micro-macro dualism' (Sibeon, ibid). For Giddens (1979: 204-5), it is 'the distinction between 'face-to-face interaction in situations of co-presence' (Sibeon, ibid) and inter-relations with other actors who are spatially/temporally 'absent' that assumes greatest importance. Put briefly….

> Giddens favours the idea of a *duality* – not a *dualism* – of action and structure. In structuration theory the 'duality of structure' is a concept which insists that agency (or action) and structure are not separate domains, but instead are 'two sides of the same coin': the notion of duality specifies that structure is not external to or apart from action – unless structure is currently being practiced (or instantiated) by people, it has no current existence (other than as 'memory traces' in people's minds). (ibid: 48-49).

Later in this Chapter, we will see how Archer is critical of what she terms 'central conflation'. Archer has Giddens's conflation of agency and structure (see above quotation) in mind (1982; 1988; 1995; 1996). Derek Layder and Nicos Mouzelis, also discussed in greater depth later in the Chapter, are also critical of approaches which conflate *agency-structure* and *micro-macro,* though Mouzelis tries to accommodate both duality and dualism in his theoretical approach. Giddensian 'methodological bracketing' emphasises a procedure that rests on a distinction between 'institutional analysis', which refers to the study of macro historical processes, and 'strategic conduct analysis' which Giddens (1984: 375) defines as 'social analysis which places in suspension institutions as socially reproduced, concentrating upon….actors' (Sibeon, ibid: 49). As Sibeon shows, Giddens argues that it is very necessary to employ such a distinction and to concentrate analysis on either agency *or* structure. Giddens's argument, in effect, is that even though in reality action and structure are not ontologically separate or different phenomena, it is in practice essential to treat them as distinct, separable phenomena. Several critics, for example Derek Layder (1984; 215; 1998b; 101) have suggested that despite arguing for a *duality* of structure at the ontological level, Giddens effectively 'lets in by the back door' an ontological *dualism* through his analytical strategy of 'methodological bracketing' (separating action and structure). Sibeon comments that ….

> the dualisms of agency-structure and of micro and macro are indispensable in social analysis; we should, as Layder (1994) aptly puts it, seek to build bridges between agency and

> structure (and between micro and macro) rather than theoretically collapse the distinctions by compressing the elements so tightly together that they cannot be separated. It follows that as well as my being critical of Giddens's insistence upon a 'duality' of action and structure, I am unconvinced by the rejection of dualism in Law's (1994) 'sociology of ordering' (which is a synthesis of structuralism, poststructuaralism, actor network theory, and, to a much lesser extent, symbolic interactionism) (Sibeon, ibid: 49-50).

Sibeon (ibid: 50) acknowledges Law's focus upon social networks (influenced by Foucauldian insights) and recursion arguably has some merits, but the author disagrees with Law's attempts to dissolve the micro-macro division (Law 1994: 11, 18, 138), and his attempt to dissolve the agency-structure distinction (Law 1994: 158-60, 103, 138). According to Sibeon, Law considers the maintenance of the agency-structure divide as 'synonymous' with essentialist/reductionist reasoning. Sibeon (ibid) considers Law to be 'profoundly mistaken' here (ibid). Law draws upon structuralist and poststructuralist material in which the 'de-centring' of subject (1994: 24) is advocated, and like Foucault, views agents as effects of discourse (ibid: 113). In Law's theorising there can be no dualism of agency and structure, because Law has wrongly assumed that the agency-structure distinction is redundant by virtue of agency being variable, contingent and relational.

There are further complications which arise from the way certain contemporary theorists deal with the micro-macro division. Nicos Mouzelis (1995: 123) contends that Anthony Giddens's rejection of the division between micro and macro is based upon a 'belief that social scientists tend to erroneously link micro with agency, and macro with structure'. This is not true of *all* social scientists, and Mouzelis argues that it is therefore unfortunate that Giddens rejects the dualism, and in doing so, abandons a useful analytical method. With *dualism*, one may examine the specific characteristics of either variable (agency, structure, micro, macro etc) rather than entertaining the possibility of missing salient points pertaining to each individual variable by engaging in an approach which favours *duality of structure* in analysis. Arguably, Giddens has tried to replace micro-macro with his own, personal conception of 'the social integration/system-integration distinction' (Sibeon, ibid: 50). In Giddensian terms, a social integration type of analysis focuses on face-to-face ('co-presence') relations between actors, and system integration analysis is concerned with 'larger' settings, such as relationships across time and space. As Sibeon (ibid) points out, Mouzelis (1995: 123) has suggested that this had led Giddens to erroneously view face-to-face interaction/relations as 'minor phenomena', whereas in Mouzelis's (1991: 83,109; 1995: 124) view (the concept of 'macro action'), such relations between powerful actors should be classed as macro-phenomena as the decisions taken by such actors at, for example, the level of head of state, may have 'repercussions that stretch widely across time-space and that may affect the lives of literally millions of people'. There are grounds for agreeing with both Giddens's and Mouzelis's refusal to link micro with agency and macro with structure, and also with Mouzelis's rebuttal of Giddens's treatment of Lockwood's (1964) social/system integration distinction. However, it is possible to challenge Mouzelis's argument to the ends that Giddens erroneously links micro with 'co-presence', suggesting that it is better to remain with the idea that face-to-face interactions are micro phenomena. Mouzelis appears to view 'co-presence' as a 'lesser' feature, rather than a defining feature of micro-interaction between actors, since interaction in specific cases involving powerful actors, such as the head of state, or of some transnational corporation, for example, often involve decisions with macro consequences in the sense that their effects may be felt across time and space. Sibeon's (ibid) view is that Mouzelis's arguments here are 'unconvincing', preferring to 'define *all* situations of co-presence – irrespective of the power of the participants – as micro-social phenomena', and adding that 'micro-social phenomena should be regarded as relatively autonomous of, though not completely detached from or unaffected by, macro-social phenomena, these being empirical questions'. The intention here, it must be noted, is not to stray too far into the vast literature upon governance, but it is useful to cite Sibeon's (ibid) example of how, if we were to apply Mouzelis's terms of governance in the European Union, 'his conception of 'macro action' (or 'macro events') would apply to, for example, interactions in the Council of Ministers or in the European Commission'. Mouzelis's notion of 'macro action' would also apply to governmental institutions such as the US Department of Energy and non-governmental (made up of 2000 scientists at 20 institutions in 6 countries) institutions such as The International Human Genome Sequencing Corporation (HGSC), were we to apply his concept to the 'public' Human Genome Project. However, Mouzelis's formulation contains some serious deficits. Firstly, the 'emergent micro-situational properties of a face-to-face situation' (an ad hoc meeting, for example) may 'significantly influence the course of interaction and thereby shape the decisions taken', and empirical examples can be found in the work of Russett and Starr, 1996: 242-3 (Sibeon, ibid). Although Mouzelis recognises interactional-situational dimensions of social action, he does not relate this to the concepts of 'macro-events' and 'macro-action'. Rather, as Sibeon (ibid) suggests, Mouzelis's references to powerful actors involved in 'macro-events', involve great emphasis upon 'positional power' (1991: 91; 1995: 24), as if he believes that personal interaction/power dynamics in the interaction and relations of 'top dogs' (such as heads of state etc) 'can have no significant emergent micro-processual or relational dimensions that affect decision-making'. Additionally, Mouzelis....

> wrongly downplays the extent to which 'macro action' involves far more than an initiating decision made by a few powerful leaders; social action and processes leading up to and during decision-making, and afterwards in the implementation phase, are often embedded in systems of power that entail a very large number of 'routinised circumstances of co-present interaction' (Giddens, 1993: 7) (Sibeon, ibid: 51-52.

In regard to these factors, it could be argued that such a hierarchical concept of macro action, which does not seem to take into consideration any notion of social chance (contingency) is problematic, and perhaps oversimplified. Thirdly, as Sibeon (ibid: 52) suggests, the events which Mouzelis (1991: 90-1) describes as 'micro-events', where decisions are taken by less 'powerful', 'micro actors' (1991: 144; 1995: 27,120) may turn out 'to have far reaching ('macro') outcomes; for example, the seeds of radical political transformation' may, in principle 'be found in encounters among 'ordinary people'', depending upon the circumstances and unintended consequences, contingency etc. Fourthly, some meetings of committees with membership of powerful actors may actually be routine, without any meaningful decision-making. Such meetings could not be labelled 'macro events', in terms of Mouzelis's formulation, 'energised by 'macro-actors'. Sibeon goes on to argue that Mouzelis's examples of 'macro events' and references to 'macro actors' 'rest on a largely systemic, role/positional conception of power' (ibid). This is in contrast to Foucauldian notions of relational power. Indeed, Sibeon (ibid) is of the view that Mouzelis

(1991: 75, 83, 90-1, 168) displays 'a tendency to adopt, in regard to the matters under discussion here, a mechanical and overly systemic view of power'. Finally, against the formulations of Mouzelis, writers such as Parsons (1995), Rhodes (1997) and Dye (1998), concerned with governance/public policy, have observed that 'big' decisions' taken by 'powerful' actors, or 'top dogs' can have only minor consequences. The work of Richardson (1996) and James (1997) contains examples of instances where technical implementation difficulties or successful resistance by implementation agents such as administrators or professionals and contingency can throw public policy off course. As stated previously, the intention here is not to review the literature on governance and public policy, but Sibeon's (ibid) application of Mouzelis's formulations to the field are worthwhile and useful illustrations of the theoretical deficits inherent in such hierarchical concepts of power. Traditional mechanisms of hierarchical, 'top down' government are in some policy sectors declining in significance in the face of non-hierarchical mechanisms of societal steering and co-governance (Pierre and Peters, 2000; Kooiman, 2003).

So far in this Chapter, we have examined the four 'cardinal sins' of *reductionism, reification, essentialism* and *functional teleology* out of which Sibeon's original framework arises as a critique. We will note that Chapter Two deals with ways of developing the 'sensitising device' further by including such concepts as (defined in the introduction and at the beginning of this Chapter) *genetic fatalism*, and the *oversocialised gaze,* as 'cardinal sins', and including a focus upon *'the biological variable'* alongside *agency-structure, micro-macro, time-space* and *power.* We have looked at metatheory (made up of meta-concepts such as agency and structure), and considered the need to adopt ontologically-flexible, post-postmodern 'sensitising devices' in the light of the inadequacies of postmodern and poststructuralist theories which entail extreme *relativism,* hyper-contextualism and idealism. We have also noted that aspects of Foucauldian notions of power are conceptually useful if we recognise the 'dialectical relationship between *agentic* and *systemic* forms of power; the *relational, contingent* and *emergent* dimensions of power, and the concept, that *contra* Foucault, aspects of power can be 'stored' in positions/roles and in social systems/networks (Mouzelis, 1995), (Owen, 2006a:193). As the framework developed here is intended to be 'applied' to the study of selected examples of human biotechnology, in this early stage of the proceedings we have examined the Human Genome Project briefly in relation to Foucauldian notions of power and in terms of non-reductionist, non-reified notions of agency-structure. Afterwards, we examined, in some depth, the meta-concepts of *agency-structure* and *micro-macro* in relation to debate around conceptual controversies and ambiguities within contemporary social theory. We looked at structuralism, poststructuralism, and discourse theory as selected examples of deterministic approaches which serve to elevate one or another conception of structure above agency. The works of Habermas and Bourdieu were also examined in relation to the same tendency. What Sibeon (2004: 56) has termed, 'miscommunication and conceptual inconclusiveness' as well as disagreement in recent, contemporary theoretical debates pertaining to *agency-structure* and *micro-macro* has been examined too, in relation to selected examples such as Mouzelis and Archer. It is clear that we need to utilise Sibeon's original framework, in the light of transitions in social theory, in a contribution towards a post-postmodern 'return to sociological theory and method', and to develop it further (which is attempted in Chapter Two). In what follows, we examine the work of Nicos Mouzelis, Margaret Archer and Derek Layder in greater depth. All three theorists are significant figures in the recent movement towards a renewal of sociological theory, and have made incisive inroads into unresolved theoretical problems (such as the relationship of agency to structure) which sociology needs to address in order to retain theoretical/empirical explanatory 'credibility' in the post-postmodern era. Comparisons between theoretical formulations are made, including between the approaches of Layder, Mouzelis and Archer's work and the concepts of Sibeon's original framework which is developed further here with which to study biotechnology.

Archer, Layder and Mouzelis: Post-Postmodern Theorists

A. Margaret Archer

Archer's greatest contribution to contemporary social theory centres upon her (1982; 1988; 1995; 1996) 'morphogenetic social theory'. Archer (1995: 1) has made it clear that, 'the "morpho" element is an acknowledgement that society has no pre-set form or preferred state' and the "genetic" part is a 'recognition that it takes its shape from, and is formed by, agents, originating from the intended and unintended consequences of their activities'. In her (1982) work, the author makes it clear that *morphogenesis* refers to the elaboration of structural forms, and *morphostasis* to their maintenance. Sibeon (2004: 97) has 'reservations' about certain aspects of Archer's framework. As we have seen, Archer's unorthodox notion of *micro-macro* has come under criticism from Sibeon (ibid.) However, the latter author values Archer's anti-conflationist formulations and dualism, adding that Archer 'has much to offer the future development of sociological theory and method'.

Part of Archer's theoretical framework involves the repudiation of what she terms, downward conflation, in other words, sociological analysis which depends upon methodological collectivism. As Sibeon (ibid) points out, in Archer's (1995) work pertaining to downward conflationist theorising, *agency* is explained *in terms of* structure. Sibeon gives the example of Talcott Parsons' later work, in which the 'central value system is said (with reference to structure and social system) to induce compatibility among indications and (in terms of agency) to socialise agents so that they energise the social system' (ibid: 97]. According to Archer (1995: 2, 16), downward conflation rests upon a less powerful 'structural conditioning' in downward conflationist paradigms actors are portrayed as unreflective, socialised beings lacking the kind of inventive capacities that might shape structure. Archer (1998: 83) observes that, in such theories, 'there is little or no sense of society as something which in its existing form may be desired by no-one'. Society, in Archer's formulations, is the 'unplanned outcome of inter-group conflict, negotiation and compromise' (ibid). Archer identifies as 'downward conflationist' the idea associated with some Marxists, structuralists, functionalists etc that society 'results from' in-built, systemic forces. Conversely, 'upward conflation' is associated by Archer (ibid) with interpretative sociology, in which society is viewed as what Sibeon (ibid: 97-98) terms, 'an aggregation of micro (face-to-face) interactions'. Archer (1995: 60) associates 'upward conflation' with interactionist sociology, with 'the neo-phenomenological school' (ibid: 84), and with methodological individualism (ibid: 34-46). As Sibeon (ibid: 98) observes of Archer....

> She argues that upward conflationist ontology wrongly treats structure as the product of agency exercised by current as distinct from previous actors, and wrongly assumes that

> structure is no more than 'other actors' and their activities. She observes that these postulates erroneously disallow the proposition that structures have aggregate and emergent properties that are 'more than' the sum of the interacting individuals and their decisions and actions, such properties having a shaping or 'conditioning' influence upon actors' forms of thought and activities (and very often, structure operates as a major constraint upon certain types of action); relatedly, upward conflation theorists tend to see structure as readily alterable by actors, providing they have the necessary motivation and information to promote social change. In both upward and downward forms of conflation, relative autonomy is withheld from agency and from structure: primacy is erroneously given to agency *or* to structure as the ultimate constituent of society, rather than investigating 'the two-way interplay between them' (Archer, 1998: 74) (Sibeon: ibid).

Archer rejects another form of theoretical reasoning, which she terms, 'central conflation' (1995: 167-8), and which she associates above all else with the structuration theory of Anthony Giddens. Archer rejects Giddensian 'duality of structure', and drawing upon the work of Bhaskar (1989a; 1989b), insists that individual social actors and society are different kinds of phenomena, rather than being 'two sides of the same coin'. As Sibeon (ibid: 98] points out, Archer (1995: 101) considers Giddens's concept of 'social practices' to be flawed in the sense that it does not 'do justice' to agency or structure, with both viewed as bound together so that there is no possibility of investigating how they interplay, or of ascertaining their respective levels of importance in a particular situation (Archer, 1998: 75, 81). Archer's analytical dualism (in which agency and structure are viewed as distinct, separable variables), suggests that temporality is central (1995: 92) to the mutually shaping relation between agency and structure.

Archer's main criticism of Giddension structuration revolves around the latter's inability to separate agency and structure in a way that would, in a sense, 'open them up' in order for us to analyse them separately, in any social context. For Archer, Giddens is an 'elisionist' (Archer, ibid: 93) who collapses agency and structure together – structure has no existence outside of its agentic instantiation.

Sibeon (ibid: 99) makes clear his respect for Archer's 'clarification of the grounds for rejecting upwards, downwards, and central conflation', viewing it as 'invaluable' for social analysis. Also viewed as 'invaluable' is her 'flexible but realist social ontology allied to her epistemological and methodological arguments; her work on a stratified conception of the actor; and her work on 'positional interests' and on the dialectics of agency and structure'. 'Positional interests' refer to vested interests. However, one aspect of Archer's framework is open to criticism. Archer's (1995: 147, 148, 253) view of the 'activity dependence' of social structure is based on the argument that 'current structures' are the effects of actions taken by actors who are, in her words, 'long dead' therefore there is a 'temporal escape' of social structures from the actions of the past. As Sibeon (ibid: 100) points out, Archer is arguing that the activity dependence of, as an example, demographic structure is 'past tense': in her words, 'here the activity dependence of such structures can be affirmed in only one acceptable way: by reference to the activities of the long dead' (Archer ibid: 143). In general terms, Archer claims that, 'we are all born into a structural and cultural context which, far from being of our making, is the unintended resultant of past interaction among the long dead' (ibid: 253). Archer's claim that present structures are effects of *past* actions is actually connected, ironically, to a confusing conflation of both the relation of present day actors to current structures, and the relation of present-day actors and structures to previous actors and previous structures. Archer appears to conflate them in a way that incorrectly downplays the extent to which present-day social contexts depend upon the activities of current actors. Arguably, Archer acknowledges that current activities in aggregate form have the effect of unintentionally reproducing society. Yet, Archer (ibid: 72), in a discussion of the activity dependence of human societies appears to suggest that current actors play no significant part in proceedings. Sibeon (ibid] argues that Archer, 'is right to refute methodological individualism as a form of upward conflation' (1995: 77; 1998: 76-8), but we must keep in mind the 'extent to which the activity dependence of current contexts or structures may relate to current activities'. Archer (1995: 145) has argued that activity dependence of social structures can be 'present time *or* past tense'. Thus, as Sibeon (ibid) points out Archer (ibid: 145) is perhaps incorrect in arguing for the existence of present-day structures that are, 'ontologically *independent* from the activities of those people here present'. As Sibeon (ibid) makes clear….

> It is legitimate to refute methodological individualism, to argue that structures have emergent properties, and to observe that the activity dependence of present structures relates to past actions; but these statements should not be taken as far as Archer's extreme claim that some current structures may be *entirely independent* of current actors. It seems that here, in a way that sits uneasily alongside some of the implications that may be thought to arise from what Archer (ibid: 141) elsewhere calls the 'no people; no society' truism, she wants to identify situations where, in her view, there is a total separation of current structure from current action.

B. Nicos Mouzelis

Mouzelis is certainly one of the leading figures in the contemporary movement towards a return to sociological theory (for example, Mouzelis, 1989; 1991; 1993a; 1993b; 1994; 1995; 1997; 2000). Mouzelis's (1991) work may be regarded as a 'landmark' (Sibeon, ibid: 101) in the movement to reject the relativistic nihilism of postmodern/poststructualist theory in favour of a, 'return to' sociological theory and method. In what follows, we examine an abridged review of selected aspects of his theoretical work. With regard, for example, to Mouzelis's extremely pertinent contributions to the 'duality/dualism' debate (covered briefly in part in this Chapter's section on 'Agency and Social Actors'), this is covered in greater detail in Chapter Three. For the time being, we will examine Mouzelis's theoretical work pertaining to the three dimensions of social action.

Mouzelis's theoretical typology contains significant treatment of agency-structure and micro-macro (1991: 196-200; 1995: 136-7), and, 'distinguishes between role/positional, dispositional, and situational-interactional dimensions of social action' (Sibeon, ibid), and the emphasis is upon focusing upon each of these dimensions in social analysis. As Sibeon (ibid) usefully makes clear, 'analysis which gives emphasis to the *role/positional* dimension is following the Durkheimian/Parsonian tradition, a form of analysis that is mainly concerned with investigation of the ways in which social positions' and generally 'predefined role scripts' part of a 'macro-institutional order' are viewed as significant 'shaping' factors in terms of local interaction. Mouzelis (1991: 106) uses the example of the 'role scripts' of a teacher and pupil to illustrate his typology, showing how, as Sibeon (ibid) acknowledges, activity and interaction in the school classroom is 'shaped' by the roles/positions of participants, at least in some cases. Mouzelis's *dispositional* dimensions of social life, borrow from Bourdieu's (1977, 1990) theory of 'habitus'. 'Habitus' refers to the general dispositions of social agents or actors, such as their skills, attitudes, values and norms, etc, that do not derive from (specific) roles such as teacher or pupil, but from wider experience of life in relation to socio-economic class, gender, ethnicity, religion etc. In theory, social actors bring to specific social encounters acquired, learned, aspects of self which though not derived from roles involved in the encounter, nevertheless partly determine the pattern of interaction that develops among the participants. In other words, *general* life experiences, 'affect how we act in any *particular* situation and the dispositional dimension helps us understand' for example, why, 'no two teachers ever perform the role of teacher in *exactly* the same way'. The third dimension of social interaction in Mouzelis's typology is the *situational-interactional* dimension, which as Sibeon (ibid) points out, is described by Mead (1967) as the emergent/contingent features of the interactional situation/encounter itself, and 'in particular to situated meanings that come into existence *during*, and as an outcome of, the process of interaction within a micro setting' (Sibeon, ibid) such as a classroom. These are emergent features of social action/interaction derived neither from positions/roles nor from dispositions.

Assessed in terms of usefulness as a post-postmodern, analytical device, Mouzelis's typology of dimensions of social action is important in terms of the development of substantive theories and the design of empirical studies.

C. Derek Layder

Derek Layder's theories emphasise the relationship between theory and social ontology in respect of methodology (for example, Layder's 1993, and 1998a work). Even when his primary concern is theory and metatheory (for example, 1994; 1997; 1998b) the question of how theory and method relate to each other runs like a thread through his theoretical work. This point is significant in any analysis of Layder's work. However, in what follows, we will concern ourselves with his metatheorising, particularly his (meta) theory of social domains. Layder, like Sibeon, Archer and Mouzelis is advocating a 'post-postmodern' renewal of sociological theory and methodology, which is ontologically flexible and avoids the 'absolutist' knowledge-claims of meta-narratives and modernist, *essentialist* paradigms. Layder 'favours a 'modest', circumscribed approach to social explanation', which retains an 'epistemological commitment to realism', recognising that 'social reality is multiform, relatively indeterminate and unpredictable' (Sibeon, ibid: 106-107). This approach is similar to Sibeon's original framework, in that it avoids unitary, *reductionist* explanations for human behaviour. As Sibeon (ibid: 107) points out, Layder's 'metatheoretical schema' draws upon some contrasting elements and he is critical of some parts of the approaches he draws upon. For example, there is his criticism of the three forms of conflation (downwards, upwards and 'central' that Margaret Archer opposes). Like Sibeon, and as is the case with the framework employed here, and in the work of Owen (2006a, 2006b, 2007a, 2007b), there is an opposition to the idea of duality of structure, indeed towards any attempt to collapse/remove important distinctions between agency and structure, micro and macro etc. Layder (1998b: 87) criticises symbolic interactionism, Giddensian structuration theory and Foucauldian insights for having what Sibeon (ibid) calls, 'a flattened ontology' which serves to ignore ''vertical' differentiation of the various spheres (or 'domains') of social reality'. In Layder's (ibid) view, differing forms of phenomena make up the social world (subjective/objective; micro/macro etc) and social 'reality' cannot be reduced to unitary, reductionist explication such as 'intersubjectivity', 'habitus', figurations' or 'social systems'. Layder appears to favour flexible, multi-factorial explanations as opposed to the relativism of postmodern/poststructuralist accounts.

As Sibeon (ibid) suggests, Layder's theory of social domains is a largely successful attempt to, 'develop a flexible, non-reductionist social ontology as a means of addressing agency-structure and micro-macro'. Layder tends to use several terms interchangeable with 'domains' and 'social domains'. These are 'dimensions', 'orders' or 'layerings', which serve to emphasise the stratified nature of his ontology. Layder contends that no domain has analytical primacy, and that the domains should be viewed as being interwoven and interdependent whilst at the same time autonomous (having distinct properties). Layder's theory of domains suggest that social 'reality' constitutes four social domains, and they relate to the subjective and objective realms. The subjective part of social life has two domains; the individual-subjective referred to as 'psycho-biography', and the intersubjective which is labelled, 'situated activity' – the objective/systemic dimension is made of up social settings and contextual resources. The latter term is explained by Sibeon (ibid) as referred to 'widespread cultural phenomena, and the distribution of resources relating to, for example, social class, gender, ethnicity, and other sources of inequalities and power differences'. Layder maintains that 'objective' (systemic) variables may influence but do not determine the subjective dimension, the latter consisting of psychobiography and situated activity. On the other hand, psychobiography (unique, asocial aspects of the person) and situated activity shape, in part, but do not actually determine social settings and contextual resources (the objective dimension of the social). As Sibeon (ibid: 108) makes clear....

> *Psychobiography* is the term that Layder uses to refer to largely unique, asocial components of an individual's dispositions, behaviour, and self-identity, these being aspects of the individual that are relatively independent of face-to-face interaction and of the macro-social sphere. It is worth noting that his concept of psychobiography (Layder 1997) in some aspects is a forerunner of the recent renewal of sociological interest in agency and the self; this interest is reflected in, for example, McNay's (2000) work on gender and agency.

Layder's emphasis upon psychobiography's relative autonomy leads him to repudiate postmodern/poststructuralist 'decentering of the subject', and the idea that social actors are the effects of discourse. Layder (1997) argues that psychobiography influences the manner in which discourses are 'handled' by individual agents. In other words, psychobiography 'mediates'. Layder is in general sympathetic to the symbolic interactionism of those such as Goffman. However, he criticises the tendency within interactionist sociology to fail to recognise the significance of psychobiography/subjectivity, which is not reducible to interaction/intersubjectivity. Layder (ibid) has criticised Elias for a failure to recognise the unique, asocial aspects of individual social actors, and Bourdieu's emphasis upon the type of dispositions (social) acquired through social group membership/experience within his theory of 'habitus' is singled out as 'incomplete' because of the lack of attention given to the psychological dispositions, pertaining to psychobiography, of individual social actors.

The second domain forming part of the subjective dimension of social 'reality' is termed *situated activity* by Layder, and refers to face-to-face social interaction as focused upon in literature of the symbolic interactionist school. As Sibeon (ibid: 108) suggests, 'Following Goffman (1983: 4), Layder (1997: 88, 93) sees situated activity in terms of situations of co-presence where two or more individuals are able to monitor and reflectively respond to unfolding action'. This would, according to Sibeon, exclude 'larger gatherings of the kind, for example, that contributed to the overthrow of Communist governments in Eastern Europe in 1989/90'. Layder (1997: 133) has suggested that situated activity subsumes 'transient', 'intermittent', and 'regularised' encounters. As Sibeon (ibid: 109) makes clear, 'Transient, one-off encounters include encounters between strangers in public places', the expression 'intermittent encounters' describes 'for example, meetings between spatially dispersed acquaintances, friends, or family members who periodically make contact with each other', and 'regularised encounters' are 'ongoing interactions among, say, family members or work colleagues'. Layder (ibid: 245) describes as a 'hotbed of creativity', the domain of situated activity which forms part of Goffman's 'interaction order'. Layder suggests that the innovative 'meanings' that frequently emerge in situations of co-presence tend to remain within the location in which they arise. That is to say that, in non-routine situations new, emergent meanings may 'travel' across time-space and have implications for the macro-social order, but in routine situations, argues Layder, the new meanings that emerge in locales generally have no implications for other locales or for the macro-social order.

Layder appears to be referring to the home, the street, workplaces etc when he uses the term *social settings* (together with *contextual resources* the term forms a dimension of the objective realm of the social). As Sibeon (ibid) suggests, in Layder's (1998a) scheme, situated activity may 'intertwine' with settings, but settings are actually relatively autonomous. Settings are variable to the extent that they are formally organised, and some settings exhibit highly formalised rules, practices, and authority relations whilst others such as family settings, are less formalised. All social settings, Layder notes (1997: 3) are 'underpinned by an elaborate social fabric of rules, understandings, objections and expectations'. It could be argued that Layder's work emphasises the importance of differentiating situated activity from social settings, the latter being *locales* of activity (Layder, ibid: 157). So, social settings may be regarded as locations of activity, relatively autonomous and of the macro-social order. As Sibeon (ibid: 109) suggests....

> In other words, within settings we encounter conditions of action-discourses, resource patterns, social positions/roles, and more or less institutionalised practices – that are inherited from the past and which, though domains interpenetrate and influence each other, are not reducible to any of the other domains.

The social domain referred to as *contextual resources* by Layder (1997: 4) comprises two elements. As Sibeon points out (2004: 110), firstly this refers to 'society wide' distributions of resources such as 'money, homes, material possessions, health care, and the like', associated with socio-economic divisions of class, gender and race; and secondly, 'widespread cultural understandings' and discourses and social practices. So, in other words, Derek Layder considers *contextual resources* to constitute a domain associated with macro-social patterns of power, domination and material inequality in addition to more widely distributed/patterned 'cultural phenomena', discourses etc.

Sibeon (ibid) makes some interesting observations about Layder's social domains theory in the light of issues pertaining to metatheory, ontological flexibility, and time-space, for example. These are all concerns which are relevant to the development of the framework here, and have been mentioned in previous discussions. Firstly, Sibeon (ibid) draws attention to the possibility of contingent links between domains. He argues that….

> While each domain is relatively independent of the others, the domains interpenetrate and overlap. There is no simple correspondence or causal connection between domains (for example, interaction in local settings may, depending on the circumstances, tend to subvert or support the macro-social order): in other words, social investigators should be alert to the development of contingent – as opposed to structurally predetermined – links between domains.

Secondly, there is the case that Layder's ontologically-flexible approach is empirically 'open' and accords no primacy in analysis towards any of the social domains. Thirdly, Layder's metatheoretical framework can be seen as advantageous in that it incorporates subjective and objective dimensions of the social. This is important because the social ontology has effectively avoided being incomplete/lopsided by this inclusion. Fourth, Layder has, according to Sibeon, effectively theorised the concepts of time-space. Therefore, as Layder (1997: 77) himself has commented, it is not possible to adequately represent his domains theory, in 'diagrammatic form'. As Sibeon (ibid: 110) explains….

We can think of social processes over time as existing in a 'horizontal' plane, but this obscures the 'vertical' layering of the four domains, and even more complexly, such layering is itself dynamic and always in 'process' (1997: 24). Also, there exist differing time frames. Interconnections of objective and subjective aspects of social reality, and of agency-structure, are complex conjunctures of time-space where the relatively short time-frames of face-to-face interaction (situated activity) meet the extended time-frames of long-standing institutional conditions that extend from the past into the present (1998b: 88).

Fifth, Sibeon (ibid) refers to Layder's stratified social ontology and 'conception of relatively autonomous but interpenetrating domains', stating that the approach contrasts with that of Nicos Mouzelis (discussed previously), who is 'close to Durkheim' when he asserts that macro-social facts must be explained in terms of other macro social facts, Layder's ontology causes him to reject the idea of explaining macro-social phenomena in terms of other macro-social phenomena, and Layder gives more emphasis to, 'the notion of a stratified social ontology in which social phenomena may legitimately be partly explained in terms of other phenomena that exist at differing levels of reality' (Sibeon, ibid: 111). Sixth, as Sibeon (ibid) shows, Layder argues that 'there is no *direct link* between action and structure'. As Sibeon correctly asserts, this relates to the theoretical debate around duality versus dualism. Layder (1997: 99, 236) suggests that there is a, 'loose coupling' between the relatively independent domains described previously. He has identified (ibid) the work of Giddens, Bourdieu, Habermas, Berger and Luckmann and Elias, for examples, as social theory which, as Sibeon (ibid) puts it, fails to, 'adequately take into account the relative autonomy of the interaction order'. Layder's view (ibid) is that system reproduction depends upon routine, and what Sibeon (ibid) refers to as 'reproductive activities in situations of co-presence'. With regard to the aspects of agency-structure discussed here, there is a general agreement between Sibeon, Archer and Layder to the extent that we need to recognise subjective/objective dimensions of society, how social conditions are dependent upon time/space and have emergent properties, and we need to incorporate these insights into social explanation.

In summary, the theorists (Archer, Mouzelis and Layder) discussed are leading theorists whose work is of importance to the development of post-postmodern sociological theory. As Sibeon (ibid: 112) suggests, in relation to Archer's morphogenetic theories...

There is considerable merit in Archer's conceptual framework, which in large part is built upon critique of three forms of conflation: downward conflation (which is associated with methodological collectivism) mistakenly attempts to explain agency in terms of structure; upward conflation, which Archer identifies with methodological individualism, interactionism, and phenomenology, is a defective form of reasoning that treats current structures as products of present as distinct from post activities, and assumes structure is no more than 'other actors' and their activities; central conflation, which Archer associates with, in particular, Giddens's theory of structuration, is rejected by Archer for its commitment to a duality of structure and action rather than a dualism in which there is a temporal separation between structure and the action that reproduces or elaborates it.

The author draws attention to other important aspects of Archer's work, including her flexible, 'realist' social ontology. However, Sibeon (ibid) views Archer's 'conception of the activity dependence of social structure' as unfortunate because it 'pushes her rejection of upward conflation 'too far', to the point of claiming that some current structures may be entirely independent of current activities.

We have examined the contribution of Nicos Mouzelis towards a sociological understanding of agency-structure and micro-macro, in particular his typology, which is influenced by Durkheim, Parsons, Bourdieu and Mead, and identifies three dimensions of social action: positional/role, dispositional, and situation-interactional. As Sibeon (ibid) suggests....

> This schema-like Mouzelis's insightful elaboration of the system and social integration distinction is developed by Mouzelis into an invaluable analytical tool. Also, its usefulness is capable of being enhanced by the inclusion of social ('organisational') actors in addition to individual human actors, and by including an explicit focus on time-space and material diffusion processes.

However, as Sibeon (ibid: 112-113) points out, Mouzelis's contribution to the theoretical debates around dualism versus duality is a mixture, on the 'positive' side of 'thought-provoking and potentially illuminating insights', and on the 'negative' side; 'flawed (meta) theoretical interpretations' such as his definition of duality as 'an unreflexive actor orientation to structure', and dualism as 'reflective actor orientation to structure' alongside retaining Giddens's concept of structure yet failing to regard dualism as a 'temporal separation between structure and action and duality as a collapsing of this separation'. Sibeon (ibid: 113) makes the interesting point that Giddens has noted Mouzelis's misunderstanding of his concept of 'duality of structure'. As Sibeon (ibid: 113) points out....

Unlike Archer and Layder, Mouzelis's thesis is that duality and dualism (as defined by him) both have a part to play in social analysis: this postulate is combined with a virtual/actual distinction that refers to paradigm and syntagm, to produce a fourfold typology that encompasses analytical categories that Mouzelis identifies as paradigmatic duality, paradigmatic dualism, syntagmatic duality, and syntagmatic dualism. These formulations by

Mouzelis, though they point to interesting questions, are ultimately unsuccessful – as they stand, they serve no useful (meta) theoretical or methodological purpose.

We have examined the contribution of Derek Layder towards our understanding of agency-structure, micro-macro and metatheory. We have also seen how two of the domains, 'psychobiography' and 'situated activity' constitute the subjective realm of the social world, and 'social settings' and 'contextual resources' refer to macro-distributions and ownership of resources, together with widespread discourses and practices. We have also noted that Layder argues against causal links/correspondence between social domains. Any 'links' are likely to be in the form of contingent 'loose couplings'. The four social domains of Layder posses lateral and vertical dimensions, and are complex conjunctions of time-space.

This chapter has outlined recent transitions in social theory which include a movement away from postmodern/poststructuralist insights towards meta-theory. We have examined the concepts and illegitimate forms of reasoning which Sibeon's original anti-reductionism focuses upon and arises out of respectively. We have examined problems with the 'cultural turn' and investigated the work of three major contributors to post-postmodern theory. It has been established that, in order to develop a 'sensitising' framework for the purpose of studying human biotechnology, Sibeon's original framework provides a viable base with considerable explanatory potential. However, as is shown in Chapter Three, it might be necessary to modify Sibeon's original framework to include a focus upon *the biological variable*; the evidence for, at least in part, biological causality in respect of some human behaviour. Evidence for this is provided from evolutionary psychology. We also discuss selected examples of attempts to 'build bridges' between the social and life sciences, for example, writers of the 'embodied' school of sociology. In Chapter Four, we examine general issues around human biotechnology, before applying the 'new' modified metatheoretical framework to the study of selected examples in Chapter Five.

Chapter Three

A Post-Postmodern Metatheoretical Framework?

In what follows, the case is made for an ontologically-flexible, metatheoretical framework for the study of human biotechnology. This would entail focusing above all upon *agency-structure, micro-macro, time-space,* modified notions of Foucauldian *power* and *the biological variable*. The previous Chapter discussed agency-structure, micro-macro and time-space (the original concepts which Sibeon's 'sensitising' framework focuses upon), and their explanatory potential. Additionally, we discussed the importance of modified notions of Foucauldian *power*, emphasising that it is important to recognise 'the *dialectical* relationship between *agentic* and *systemic* forms of power; the *relational*, *contingent* and *emergent* dimensions of power, and the concept that *contra* Foucault, aspects of power can be 'stored' in positions/roles and in social systems/networks' (Mouzelis, 1995) (Owen, 2006a: 193). This Chapter includes arguments for the modification of Sibeon's original framework to include a focus upon *the biological variable*; the evidence from evolutionary psychology and behavioural science for, an at least in part, biological basis for some human behaviour. The Chapter also explores further ideas pertaining to dualism between sociology and biology as opposed to duality, and looks at the attempts of those who seek to redefine the relationship between 'nature' and 'the social'. This includes writers of the 'embodied' school. The argument for the inclusion of 'extra' cardinal sins (alongside Sibeon's original *anti-reductionist* concepts of *reductionism, reification, essentialism* and *functional teleology*) such as '*the oversocialised gaze'* and *genetic fatalism'* is also reinforced in this Chapter. In what follows, we briefly codify the framework which is employed in Chapter Four, or rather 'applied' to selected aspects of human biotechnology. Afterwards, we begin to look at the possibilities of 'building bridges' between the social and life sciences, and at the possible need to acknowledge *the biological variable*, neglected by Sibeon's original framework.

Codification of the Metatheoretical Framework

The previous Chapter made the case for the post-postmodern explanatory potential of *metatheory*. Briefly, as Sibeon (2004: 13) makes clear, *metatheory* is designed to equip us with 'a general sense of the kinds of things that exist in the social world, and with ways of thinking about the question of how we might 'know that world'. The framework employed here is an example of *metatheory* consisting of methodological as distinct from substantive generalisations. It is argued here that the framework is capable of contributing both to the 'return to' sociological theory and method, and to a post-postmodern sociology of human biotechnology. The 'pay off' will hopefully lie in the framework's explanatory potential in terms of theoretical development. Rather than being 'hot house debate' it is the author's contention that the framework's usefulness lies in its avoidance of anti-foundational *relativism, oversocialised* (harshly 'environmentalist') accounts; the tendency towards *genetic fatalism* (the equation of genetic predisposition with inevitability); and its adoption of the 'sociological realism' of those such as Mouzelis (1991, 1993, 1995) who seek a 'return to' sociological theory and methodology. The framework has been employed by Owen (2006a) in a theoretical contribution towards a 'Post-Foucauldian' sociology of aging, by Owen (2006b) in the form of 'Genetic-Social Science' in order to study biotechnology, and by Powell and Owen (2005) to critique the 'Bio-medical model' of aging.

The 'sensitising device' arises out of a critique of six forms of illegitimate theoretical reasoning. The first four are found in Sibeon's (1996) original framework (which he called *anti-reductionist sociology*). They are, *reductionism, essentialism, reification* and *functional teleology*. To recap, a *reductionist* theory is a theory that attempts to explain social 'reality' in terms of a single, unifying principle such as 'patriarchy' (Hindess 1986, 1988). *Essentialism*, 'presupposes in aprioristic fashion a necessary unitariness' or 'homogeneity' of social phenomena' (Sibeon 1997b: 1). *Reification,* 'involves the illegitimate attribution of agency to entities that are not actors or agents' (ibid). *Functional teleology*, refers to 'illicit attempts to explain the causes of social phenomena in terms of their effects' (Betts 1986: 51). The second two, which are employed by Owen (ibid) and Powell and Owen (ibid) and here, are *Genetic Fatalism* and the *Oversocialised Gaze. Genetic Fatalism* refers to the widespread tendency within social science to associate genetic predisposition with inevitability (Ridley, 1999, 2003). *The Oversocialised Gaze* refers to 'Environmentalist', accounts characterised by a strong antipathy towards genetic or even partially genetic explanations, such as Giddens's (1993: 57) suggestion that, 'human beings have no *instincts* in the sense of complex patterns of unlearned behaviour'. Aside from focusing upon *agency-structure*, *micro-macro* and *time-space* (included in Sibeon's original framework and explored/defined in the previous Chapter), the 'new' framework also entails focusing upon the *biological variable*; the evidence from evolutionary psychology and behavioural science for, an at least in part, genetic basis for some human behaviour (Ridley, 1999, 2003).

'Building Bridges'

In what follows, we examine the work of several authors who have attempted to straddle the biology-sociology divide and consider whether it is possible to move beyond the established 'starting point' for a *biological sociology* outlined in the work of Shilling (1993). Among authors who have attempted to 'cross the divide', between the social and life sciences, Layder's (1993, 1997) work incorporates the interesting and useful concept of *psychobiography* (discussed in terms of explanatory potential in the previous Chapter), the largely unique, asocial components of disposition, behaviour, and self-identity. These aspects are regarded by Layder as relatively independent of face to face interaction. For Layder, human beings are composed of unique elements of cognition, emotion and behaviour that are, in some sense, separable from the social world, while at the same time related in various ways to social conditions and social experiences. Layder is critical of postmodern attempts to 'de-centre' the subject, 'the cultural turn' in general, and Symbolic Interactionism. The latter school of sociology is singled out for particular criticism because of its tendency to emphasise intersubjectivity to an extent that largely ignores individual subjectivity (or *psychobiography*).

There are possibly philosophical and sociological connections between the psychobiographical components of Layder's 'Social Domains' theories and the work of Benton (1991, 1993, 1994, 1999, 2003), and Hochschild (1983, 1990). Arguably, all three authors have managed to straddle the biology-sociology divide in a way that so many sociological accounts have failed to do. Hochschild's work on emotion, as is the case with the work of Taylor (2006) stands in sharp contrast to 'social constructionist' models of emotion. Such 'mainstream' sociological accounts tend to prioritise the biological and the social respectively. Hochschild (1983) adopts a contrasting position which occupies an analytical space between the two extremes. For Hochschild, emotion is unique among the human senses, related to action and cognition. This is similar to Layder's views pertaining to unique, psychobiographical aspects of the person. Emotion emerges, according to Hochschild, when bodily sensations are joined with what we see or imagine, and on this basis we discover our own particular views of the world, and our readiness to act within it. Such an approach serves as a valuable lesson on the role of *the biological variable* in sociological explication. For a considerable time, what Freese et al (2003) call 'biophobia', a hostile reaction on behalf of sociologists towards explanations which evoke biology has been in evidence. The mere mention of biological causality has tended to lead to accusations from 'mainstream' sociology of a vulgar and crude reductionism. As Hochschild (ibid) argues, while a sociology of emotions needs to 'go beyond' the biological, this does not mean leaving it out altogether. Conversely, if we choose to incorporate biology into sociological explanation, this does not necessitate an overly deterministic role for the latter. Cases in point are the work of Owen (2006a, 2006b), Powell and Owen (2005) and Quilley and Loyal (2005). In the latter example, it is argued that Elias's work provides a 'central theory', serving to integrate diverse theoretical traditions within sociology, whilst providing a framework for establishing a synthesis across the full range of social and biological human sciences. Any sociological enquiry into human emotions will surely confront 'the limits of the social', but it is the case that human biology displays a degree of plasticity in relation to wider socio-cultural influences (Ridley, 1999: 306). For example, feelings are 'not stored' inside human social actors, but the management of feeling may contribute to the creation of it. Here Hochschild, has, arguably, developed a fairly sophisticated perspective on human emotions which side-steps crude, reductionist debates.

Benton's (1991, 1994, 1999) work criticises the 'idealist' tendency in contemporary social theory and the postmodern 'cultural turn'. He suggests that extreme relativism in the philosophy and sociology of science has interbred with a 'pervasive misreading of the Saussurean thesis of the arbitrariness of the relation between sign and signified' and also that 'Nietzschean moral and epistemological nihilism' has produced luxuriant growths of hyper-idealism' (Benton 1994: 45). The author also suggests that the complex network of, 'categorical oppositions (mind/body; culture/nature: society/biology, meaning/cause, human/animal) which structure the discourses and define the research agenda of these prevailing traditions now constitute intellectual obstacles' in the way of meeting the challenge to 'settled' conceptual forms in 'mainstream' sociology (ibid: 25). Benton considers that his work has provided a preliminary blueprint for how 'dualistic', anti-naturalistic programmes in the contemporary human sciences should be understood. As it is made clear in the previous Chapter, the framework employed here favours *dualism* as opposed to a d*uality of structure*. As it is explained later in this Chapter, this is in order to avoid the compacting of not only agency and structure, but also to avoid the collapsing of distinctions between the 'biological' and 'the social'. According to Benton (ibid), 'dualistic' programmes are best understood as primarily defensive reactions to 'reductionist' disciplines such as Social Darwinism. The author's view is that the 'task' for any re-alignment of sociology and related social sciences with the life sciences is one of providing conceptual room for organic, bodily and environmental aspects, and aspects of social life, to be given their 'proper place' without discarding the achievements of classical social science in defense of the autonomy of those disciplines vis-à-vis the life-science specialisms. Importantly, like Hochschild, Benton (ibid) suggests that a realignment or even integration between disciplinary fields such as biology and sociology does not have to mean reductionism, or a reduction of the one to the other.

Recently as Owen (2006a) suggests, Armstrong (1987: 1217), drawing upon Foucauldian insights, has warned against too readily endorsing 'an alliance of bio-social perspectives'. Similarly, a range of Foucauldian arguments have emerged, claiming that medical power should be seen as a 'dangerous' form of power and surveillance (Katz 1997; Biggs and Powell 1999; Powell 1998; Powell and Biggs 2000). These perspectives go some way towards offering *social* understandings of disease that may be juxtaposed against the biological definitions operant in hospitals etc. Scepticism towards biological 'determinism' can also be found in the Foucauldian analysis of Powell and Biggs (2004: 2) in which it is argued....

> A significant contribution of sociology as a discipline has been to highlight how individual lives and illnesses which were thought to be determined solely by biological, medical and psychological factors, are, in fact heavily influenced by social environments in which people live.

The authors are arguably correct to suggest that social variables play a large part in 'influencing' behaviour, lives and illnesses. As Ridley (1999: 306) suggests, social variables are probably in total 'more important than genes in nearly all behaviours'. However, it is argued by Owen (2006a, 2006b, 2007a, 2007b) that we should avoid associating genetic predisposition with inevitability. As previously stated, the framework employed here arises out of a critique of several, illegitimate forms of reasoning, one of which is *genetic fatalism*. The tendency to equate biological predisposition with inevitability is fairly widespread in social science, according to Ridley (ibid).

Newton's (2003: 35) work pertaining to the possibility of 'marriage' between sociology and biology, suggests that the human body can be 'both extra-discursive and deeply embedded in the social fabric' as can be observed in the human smile, body scripts of birth, maturation and death etc. In his view, such examples serve to illustrate the *social* reliance of the 'extra-discursive body' and how any sociological account of the social world is deficient if it, 'ignores the fact that human beings have *biological* bodies' (ibid). In Newton's view, the significance of the human body for the social serves to emphasise the need to formulate concepts capable of integrating the two. However, he highlights some of the problems facing those such as Benton (1991), Bury (1995, 1997), Williams (1998) etc who are attempting to cross the 'Great Divide' in the sense of furthering the project of a material-corporeal sociology. In particular, Newton draws attention to the reliance on animal studies in order to explain human health (Freund 1998: 854-5; 1990: 464, 1998: 277) and Williams (1998: 129, 2001: 71), arguing that, 'reference to animal studies incites a reductionist account of human health and appropriates rather inappropriate analytic concepts-features which appear to contradict Freund and Williams' expressed concerns' (ibid: 28). Newton sympathises with Shilling's (1993) attempts to analyse the human body as *simultaneously* biological and social, viewing this 'Third Way' as a starting point for a *biological sociology* which goes beyond naturalistic and social constructionist views of the human body, while retaining some of their useful insights. However, Newton illustrates how there are serious methodological and epistemological problems facing current attempts to move beyond Shilling's (ibid) starting point. He suggests that a strong desire to counter constructionism may lead to an 'insensitivity' towards constructionist arguments, citing the work of Freund and Williams again as examples of authors who lack a, 'reflexivity to their own *emotional* evocation of the supposed links between the social (eg inequality), the emotional (and its physiology) and ill-health' (Newton ibid: 36).

In sum, Newton urges us to remember that we remain 'at the starting point' of a marriage between the social and the biological. In his view, we cannot rely upon current psycho-physiological research in order to conceptualise biology, and we face further problems in relation to the complexity of the relationship between the human body and the social world. Indeed, in his view, we cannot necessarily interpret a particular physiological measure as a 'response' to social stimuli. For Newton, 'neither the social nor the biological generally operates in this singular fashion' (ibid). He also draws attention to the difficulties of isolating 'singular relationships' within the biological field, characterised as it is by great complexity.

It is argued by Owen (2006a, 2006b, 2007a) that we *can* move beyond Shilling's (ibid) 'starting point' for a marriage between biology and sociology. We can incorporate the concept of *genetic fatalism* (as a 'cardinal sin') into a 'new' framework, and focus upon *biological variables* (evidence for genetic causality). Arguably, the greatest error in the work of many of the authors who have attempted to straddle the biology-sociology divide (in particular, Benton 1991, 2003; Shilling, ibid; and Newton ibid) appears to lie in the tendency not to equate predisposition with genes but to equate predisposition with *inevitability*. As Ridley (1999: 307) clearly states....

> Suppose you are ill, but you reason that there is no point in calling the doctor because either you will recover, or you won't: in either case, a doctor is superfluous. But this overlooks the possibility that your recovery or lack thereof could be caused by your calling the doctor, or failure to do so. It follows that determinism implies nothing about what you can or cannot do. Determinism looks backwards to the causes of the present state, not forward to the consequences.

In the 1970s, after the publication of E.O. Wilson's book *Sociobiology,* there was a counter-attack against the idea of genetic influences upon human behaviour, led by Richard Lewontin and Stephen J. Gould (Wilson's Harvard colleagues). Their dogmatic slogan: 'Not in our genes!' may have been a plausible hypothesis at the time, but as Ridley (ibid: 306) argues, 'After 25 years of studies in behavioural genetics, that view is no longer tolerable. Genes do influence behaviour'. Perhaps we may move beyond Shilling's (ibid) suggestion that the human body is *simultaneously* biological and social, by acknowledging Ridley's (ibid) point about genetic influences. We should perhaps recognise that, even after the genetic discoveries, Ridley mentions, environmental influences remain enormously important. In the light of Newton's (ibid) call for a, 'more sophisticated' understanding of biology and psychology, perhaps we should strive to avoid sociological accounts which deny the possibility that the physical can be 'at the mercy' of the social, such as those of Armstrong (ibid). As Ridley (ibid: 153) suggests….

> If genes are involved in behaviour then it is they that are the cause and they that are deemed immutable. This is a mistake made not just by genetic determinists, but by their vociferous opponents, the people who say behaviour is 'not in the genes', the people who deplore the fatalism and predestination implied, they say, by behaviour genetics. They give too much ground to their opponents by allowing this assumption to stand, for they tacitly admit that if genes are involved at all, then they are at the top of the hierarchy. They forget that genes need to be switched on, and external events – or free-willed behaviour – can switch on genes.

There is a growing, emerging literature on the subject of the possibility of a 'marriage' between the biological and social sciences. So far we have examined the work of several theorists of the 'embodied' school, and a 'voice' from evolutionary psychology in the form of Ridley. It is important to widen and deepen the focus to include an examination of more diverse sources, and to tentatively apply some of the concepts from the 'new' framework to some of the theoretical positions in the debate (for example, the 'embodied school'). In Chapter Three, to recap, we examine general issues around human biotechnology, perspectives from sociologies of science and look at the Human Genome Project more closely than we did in Chapter One. In particular, we examine the sociological implications of the Project. Chapter Four deals with a more in-depth application of the framework to the study of selected examples of human biotechnology. For the moment, we will examine further the literature pertaining to links between sociology and biology, and consider whether it is viable, given the evidence, to include the *biological variable* alongside *agency-structure, micro-macro* and *time-space*.

Williams (2003: 550), another writer associated with the 'embodied school', or 'the sociology of the body' makes clear his position within the debates around the possibility of 'marriage' between the social and the biological, and the broader context within which the debates are located. The author refers to the current 'state' of the debates within sociology, and describes several 'schools of thought'. Alongside those such as Dickens (2000) who attempt to develop an 'alternative kind of social Darwinism' (Williams ibid: 556), there are those such as Williams himself, Freund, Benton and Shilling who may be located within 'embodied sociology'. In Williams' view, it is the 'embodied' theorists who have been particularly important in, 'challenging if not dispensing with dualist legacies of the mind/body, reason/emotion, biology/society kind' (Williams ibid: 550). Also referred to by the author are the recent developments within *the sociology of health inequalities*, such as the 'psychosocial perspective' of Wilkinson (1996, 2000a, 2000b) which represents something of a departure from 'embodied sociology'. As Williams (ibid) explains, for Wilkinson (ibid), health is 'no longer limited primarily by the direct effects of material factors in rich societies that have gone through the 'epidemiological transition''. Elstad (1998: 600) has commented that there are three central contentions with the *psychosocial* approach, which are, '(1) the distribution of psychological stress is an important determinant of health inequalities in present-day affluent societies, (2) psychological stress is strongly influenced by the quality of social and interpersonal relations, and (3) the latter are determined to a large extent by *the magnitude of society's inequalities'*. As Williams (ibid: 552) points out, there are four main sources for the psychosocial perspectives of Wilkinson (2000a). These are the *'social stress'* approach, the related *'self-efficacy'* approach, the newer *'sociology of emotion'*, and the *'social cohesion'* approach.

For Elstad (ibid: 601) it is not a case of, 'whether *some* health-related bodily changes may follow from mental appreciations of external circumstances'. In the author's view, recent reviews leave, *'few doubts that psychological stress, generated by despairing circumstances, insurmountable tasks, or lack of social support, can influence disease-related patterns'*. The author identifies examples such as, 'the endocrine system (eg secretions of catecholamine and cortisol), and the immune system (eg secretions of catecholamine and cortisol), and the immune system (eg number of T-cells and Natural Killer cells)'. However, the problem is, *'to demonstrate that such changes are large enough and long-term enough to affect health in a significant way'*. Newton (2003) raises likewise concerns, as do Wainwright and Calnan (2002). However, Williams (ibid: 553) considers the idea that we should, *'question the whole relationship'*, premature as 'the pathways are *more than* plausible'.

Williams (ibid: 555) discusses the 'heated debate' that has been sparked by 'neo-material critiques of psychosocial perspectives or pathways, including important questions concerning the 'causes of the causes', the deeper relations of (class) inequalities and the (global) role of neo-liberalism (Coburn, 2000a, 2000b; Scambler 2000; Wilkinson 2000b)'. In Williams' view, what is required, is, 'the need to think in terms of relationships between, 'social structure, ecological conditions of life, bodily constitution and activity, social and cultural processes, and mental and emotional dispositions, skills and experiences' (Benton 2003: 285). Williams considers that such a position avoids what Benton (ibid: 293) has called, 'the simplistic opposition of 'nature' versus 'nurture' in relation to health'.

Interestingly, on the one hand, Williams (ibid: 556) argues that, 'if we want to challenge dualism on all fronts, as 'material corporeal sociologists' appear to, then one important route, as Birke (2003) reminds us, must be to take on board the growing calls to recognise at least some kinds of non-human animals as 'clever, adaptable, aware and cultural – just like us''. On the other hand the author appears to condone dualism when he states, *'we need to respect the discrete analytic potential (and autonomy) of various disciplines*, biology and sociology included, thereby countering the (reductionist) slide from the having of *something* important to say about the world and our place within it, to the having of *everything* to say on these matters' (ibid: 558). Despite this apparent contradiction in terms, Williams sees 'the way forward' in the biology-society debate as a 'multi-disciplinary' approach which would serve to 'dissolve the dualistic opposition between 'society' and 'nature' without giving in to either a social constructionist reduction of nature to culture, or the reverse reduction of social life to a mediated epiphenomenon of the human genome (Benton, 2003:292)'.

EVOLUTIONARY PSYCHOLOGY AND 'NATURE VIA NURTURE'

Evidence for the idea that human beings evolved complex behaviour by the addition of instincts to those of their ancestors rather than by replacing instincts with learned patterns of behaviour can be found in the work of the psycho-linguist, Pinker (1994). As the work of Owen (2006b) shows, Pinker's argument is that the intention of most social scientists has been, and is, to examine and trace the ways in which behaviour is influenced by the social environment. We should perhaps examine instead the ways in which the social environment is produced by our innate social instincts. As Ridley (1999, 2003) has suggested, the 'fact' that all human beings smile at happiness and frown when worried may be expressions of unlearned behaviour (instinct) rather than culture.

The evolutionary psychologists Barkow et al (1992) suggest that culture is the product of individual human psychology more than vice-versa. Arguably, it has been a major mistake to oppose 'nature' to 'nurture', because learning is, according to the authors, dependant upon innate, instinctive capacities to learn, and there are innate constraints upon what can be learned. As Ridley (1999: 103) makes clear, the 'evolutionary' in evolutionary psychology refers not to natural selection itself but to the concept of adaptation. Biological organs can be 'reverse engineered to discern what they are 'designed' to do', in the same way as machines can be so studied. Pinker (ibid) emphasises that machines are *meaningless* except when described in terms of their functions. In the same way it is meaningless to describe the human or animal eye without mentioning that it is 'designed' for the making of images. Barkow et al (ibid) argue that the same principles apply to the human brain, and its modules are likely to have been 'designed' for particular functions. Ironically, as Ridley (ibid) points out, 'it was the argument from design that kept evolutionary ideas at bay throughout the first half of the nineteenth century'.

'The evidence that grammar is innate is overwhelming and diverse', claims Ridley (ibid: 104). The author refers to the language instinct possessed by all human beings as a complex and sophisticated adaptation for communication. He conceives how 'it was of great advantage to our African ancestors to share precise and complex information at a level unavailable to any other species'. Ridley considers the evidence that grammar is innate rather than learned to be abundant….

> The evidence that a gene somewhere on chromosome seven usually plays a part in building that instinct in the developing foetus's brain is good, though we have no idea how large a part that gene plays (ibid).

However, many social scientists appear to be firmly resistant to the idea of human instincts. For example, Giddens (1993: 57) suggests that 'Human beings have no *instincts* in the sense of complex patterns of unlearned behaviour'. Yet, as Owen (2006b:96) points out, 'the evidence for instinct, and for genes whose effect seems to be to achieve the direct development of grammar remain strong'. Certainly, the work of Dunbar (2003a, 2003b, 2004), Barrett, Dunbar and Lycett (2002) and Dunbar, Knight and Power (1999) seems to support this position. As Ridley argues (ibid: 105), it appears to be the case that many social scientists prefer to think, contrary to an abundance of biological evidence, that genetic influences upon human language are, 'mere side-effects of its direct effect on the ability of the brain to understand speech'. Arguably, the 20th century paradigm to the ends that only non-human animals possess instincts begins to develop cracks in the foundations, 'once you consider the Jamesian idea that some instincts cannot develop without learnt, outside inputs' (Ridley ibid). The reference here is to William James, the early psychologist who believed that human behaviour revealed evidence of more separate rather than fewer instincts than non-human animal behaviour. Yet the belief that human beings possess no innate characteristics outside of, 'a set of simple reflexes, plus a range of organic needs' (Giddens, ibid), appears to be firmly entrenched within the discipline of sociology in particular.

Despite reservations about Lewontin's (2000) *genetic fatalism*, Owen (2006b: 905), 'recognises that the author has cogently identified biological discourses which are remarkably similar to the dominant ideas of their day'. This point is emphasised because Lewontin is a most prominent example of sociologists who appear to deny biological causality. Lewontin's contention that 19th century 'racist scientists' produced 'racist science' is controversial to say the least, and must be taken seriously. His suggestion that a 'democracy of the elite' operates within science is far from a trifling issue. The implications are that the apparatus of knowledge called 'scientific knowledge', depending upon elaborate metaphors, is available only to a restricted elite. Additionally, his suggestion that there is a 'dark side' to the manner in which the Creationist movement in the US Bible Belt has 'thrown out' all science rather than re-evaluate inherited notions of God, importantly identifies a mindset which automatically opposes any form of change to the mythical 'timeless' landscape (Owen, ibid). Likewise, Lewontin's timely criticisms of the idea of the Human Genome Project as a 'recipe book' for life, cannot be dismissed lightly.

Particularly strong criticisms of evolutionary psychology can be found in the work of Midgley (2000). The author argues that evolutionary psychologists 'teach' determinism and nihilism, whilst treating human beings as solitary entities rather than social animals. Daly and Wilson's (1998) work appears to contradict Midgley's point regarding the reliance upon social atomism by evolutionary psychologists. The authors examine step-relationships, and acknowledge that they 'exist in all societies' (Daly and Wilson ibid: 37-8).

Perhaps at this point, it might be useful to locate evolutionary/psychology within wider sociological debates around 'the new genetics'. David (2002: 303) locates evolutionary psychology's 'mass modernity account of human intelligence' within discourses pertaining to the 'naturalisation of society'. For David, current sociological interest in genetics and biotechnology 'focuses attention on three interrelated fields: the 'naturalisation of society' (Benton, 1999; Dupre, 2001); the 'socialisation of nature' (Dickens, 1996; Irwin, 2001); and the 'normalisation of the individual' (Ettorre, 2000; Petersen and Bunton 2002)' (ibid). The 'socialisation of nature' refers to 'attempts to apply genetic technologies to modify non-human nature (plants and animals)' (ibid). The 'normalisation of the individual' focuses upon, 'the application of genetic diagnostics and therapies within the realm of human health and medical practice' (ibid). General overviews of evolutionary psychology (such as those provided by Dunbar, Barrett and Lycett 2006; Barkow, Cosmides et al, 1992; Pinker 1997; Buss 1999) tend to emphasise that evolutionary psychology 'argues that the mind is a collection of special-purpose software designed by natural selection to solve the problems of survival and reproduction that faced our ancestors' (Curry, 2003: 1).

As previously emphasised, evolutionary psychology is not without its critics, and our task after examining the literature is to decide whether there is sufficient evidence from the discipline to justify modifying Sibeon's original framework so that it entails focusing upon *biological variables* (evidence for biological/genetic causality, at least in part, with regard to some human behaviour).

David (ibid: 304) considers evolutionary psychology to be, 'only the latest manifestation of biological reductionism', seeking to 'identify the roots of the contemporary human condition in the stone age/s'. David is sceptical about 'reverse engineering' (the attempt to 'infer the ancestral conditions, called the environment of evolutionary adaptation, that would have made certain genetically inherited behaviour-inducing modules increase their bearer's reproductive success') (ibid), because it is 'empirically untestable' (ibid: 312). The author identifies the 'existence of hard-wired mental 'modules'' designed to produce 'programmed response'. These 'adaptionist' principles imply that 'what exists is said to have evolved' (ibid: 304). While others such as Rose and Rose (2000) have criticised evolutionary psychology regarding theoretical deficits pertaining to rape, social competition etc, David concentrates upon addressing the 'core proposition' of the discipline, the 'modularity of mind'. The author feels that sociology must 'challenge reductionist science', of which evolutionary psychology is an example, whilst contributing to a wider understanding of, 'what it does mean to be human' (ibid). David identifies the concept of the modularity of the human mind as a central concern in the work of several leading evolutionary psychologists, such as Barkow et al (1992), Pinker (1995, 1999) and Gigerenzer et al (1999). The basic idea is that the mind is 'made up' of a series of specific modules adapted to provide 'functional responses' (David, ibid) to specific stimuli/problems, rather than being a 'single processor able to use general reasoning principles or mechanisms to solve problems' (ibid). Therefore, in David's view, this means that evolutionary psychologists may apply the principles of natural selection to psychological phenomena. The module, 'by analogy with the gene' (ibid), becomes the unit of selection. Referring to Fodor's (1993) formulation of modularity, which assumes a 'dual processor model', David makes the point that 'far more of our *thinking* is 'pre-programmed' than is assumed in dual processing theory'. However, the evidence for this claim appears to be lacking. The author argues that the general programme of evolutionary psychology is to 'extend the logic of 'hard-wired' mass modularity to every aspect of human behaviour' (ibid).

Referring to the common metaphor given for evolutionary psychology's concept of mass modularity, the Swiss Army Knife of Barkow et al (1992), David (ibid: 305) considers such an account (in which the human brain/mind is conceptualised as a 'multiple of relatively independent blades, each one built to carry out a specific task') to be a 'purpose-built intellectual module serving a singular function', which is to deny the 'possibility that humans could be other than they are'. The author adds that, 'such a model of the mind fits very nicely into the current intellectual and social climate'. He goes on to quote Hilary Rose (2000: 107) who argues that, 'a restored and re-energised neoliberalism has provided the perfect ecological niche for a new wave of biology as destiny'. Interestingly, Curry (2003: 9) in a review of Hilary Rose and Steven Rose's (2000) *Alas, Poor Darwin: Arguments against evolutionary psychology*, suggests that the Roses have not provided evidence that the ideology is motivated by 'unexamined political presuppositions'. David's (ibid) claim to the ends that a neatness of fit between the model of the mind favoured by evolutionary psychologists and current intellectual and social fashions enables the outlook to flourish is arguably contestable. David (ibid) refers to evolutionary psychology advocates (though he does not provide an example) who suggest that, 'rationality cannot be defined outside its reproductive functionality', and 'if a way of *thinking* helps an organism survive and reproduce it is said to be rational'. As evolutionary psychology is well adapted in the political environment it, therefore displays evolutionary rationality according to the author. David (ibid) goes as far as to say that, in its own terms EP therefore displays evolutionary rationality. Aside from the lack of a referenced example, David arguably engages in *reification* here (one of the 'cardinal sins' of the framework), and refers to evolutionary psychology as if it is a social actor possessing the agency to formulate and act upon decisions.

David (ibid) does, however, provide some evidence from the work of Karmiloff-Smith (1992, 2000) which may demonstrate some weaknesses in the theory of mass modularity. According to David, Karmiloff-Smith (ibid) provides evidence that studies of brain damage taken by evolutionary psychologists to demonstrate modularity, actually demonstrate the '*plasticity* of the brain' (David: ibid). Damaged brains show forms of wiring which are different from those of non-damaged brains and not only in 'the dysfunctional areas' (ibid). There is evidence, it is claimed, that the human brain 'appears to compensate for damaged regions by creating alternative neural connections' (ibid). As David suggests, 'far from proving full modularity, such damage limitation suggests that, while there is a degree of modularity, the brain actively modifies any blueprint it may be built upon to limit general dysfunction' (ibid). Essentially, David appears to be suggesting that Karmiloff-Smith's work contains evidence that there is a less 'fixed' relationship between 'region' and 'function' within the human brain than is implied by the evolutionary psychological notion of modularity. David (ibid: 304) refers to Fodor's (1983) original formulation of mass modularity, and has, as we have seen, used Karmiloff-Smith's (1992, 2000) work to identify possible deficits in the theory. However, Fodor (2001) published *The Mind Doesn't Work That Way*, which effectively dismantles his original formulation of mass modularity. There is no mention of Fodor's later work in David's (2002) paper. To be fair to Fodor (ibid), Ridley (ibid: 66) shows how the author argues that 'breaking down the mind into separate computational modules was by far the best theory around' (and David (ibid) does acknowledge 'a degree of modularity'), 'it did not and could not explain how the mind works'. According to Ridley (ibid) Fodor (2001) is arguing that….

> Minds are capable of abducting global inferences from the information supplied by the parts of the brain. You may see, feel and hear raindrops with three different brain modules linked to different senses, but somewhere in your brain resides the inference: 'it is raining'. In some inevitable sense, then, thinking is a general activity that integrates vision, language, empathy and other modules: mechanisms that operate as modules presuppose mechanisms that don't. And almost nothing is known about the mechanisms that are not modular. Fodor's conclusion was to remind scientists just how much ignorance they had discovered: they had merely thrown some light on how much dark there was.

David (ibid: 309) refers to 'genetic reductionist accounts of the mind' which draw upon 'misleading analogies with computers'. Citing the work of Pinker (1999), David refers to the tendency, as he perceives it, in evolutionary psychology to view intelligence as 'computationally modular'. This model is challenged by those in theoretical sociology such as Collins and Kusch (1998), and those in neuroscience such as Van Gelder (1995). Collins and Kusch (ibid) are said by David (ibid) to have developed an analytical framework for distinguishing between types of action. In theory, the frame enables greater insight into differences between the capabilities of human beings and machines such as computers. The authors highlight the 'higher-order intelligence which most human 'work' requires, intelligence which machines are still capable of achieving' (David: ibid). Collins and Kusch (ibid) distinguish between action and behaviour, with action seen as 'intentional' rather than reflex/mechanical. Within the action category, as David (ibid) shows, Collins and Kusch (ibid) distinguish between polimorphic and mimeomorphic forms of action. 'Mimeomorphic action is action that can be specified fully in terms of its behavioural components (and the situations in which it is appropriate) without an understanding of the 'form of life' of which that action is a part' (David: ibid). These then, are intentional actions. Polimorphic actions are 'those where the capacity to successfully follow the rules of the form of life in which the action takes place is crucial for the correct performance of the action' (ibid). Collins and Kusch (ibid) give the examples of 'writing love letters' and 'voting' as acts which cannot be mechanised, as according to David's (ibid: 310) account of their work, 'the act is not the mechanics, but the expression of an intention requiring the capacity to participate in, at however simple a level, the political life of a community'. Polimorphic actions are 'higher-order' forms of action 'requiring the capacity to follow cultural rules that are themselves context-dependent and which, therefore, cannot be specified outside the cultural practices of which they are a part' (David: ibid). For Collins and Kusch, forms of 'playful' polimorphic action include human attempts to 'enact the same behaviour to relieve boredom or express individuality' (David, ibid) as here intention rather than the capacity to mimic behaviour is manifested. 'It is polimorphic also to the extent that rule bending is itself a social institution with culturally enacted rules that must be interpreted correctly, in the eyes of other cultural players, to be carried out successfully' suggests David (ibid). Collins and Kusch (ibid) are, according to David (ibid) firmly in favour of the concept that polimorphic actions are not modular in themselves because the rules they enact are unable to be specified in a modular way. David's critique of evolutionary psychology appears to treat evolutionary psychology as 'only the latest manifestation of biological reductionism' (ibid: 304). However, as Ridley (2003: 66) shows, within the 'school of thought', there are subtleties which those such as David (2002) appear to miss. For example, with reference to computation in the human brain, Ridley (ibid: 66) suggests that….

To build a brain with instinctive abilities, the Genome Ordering Device lays down separate circuits with suitable internal patterns that allow them to carry out suitable computations, then links them with appropriate inputs from the senses. In the case of a digger wasp or a cuckoo, such modules may have to 'get their behaviour right' first time and may be comparatively indifferent to experience. But in the case of the human mind, almost all such instinctive modules are designed to be modified by experience. Some adapt continuously throughout life, some change rapidly with experience then set like cement. A few just develop to their own timetable.

Far from being the 'genetic reductionist accounts' which David (ibid: 309) refers to, it would appear that evolutionary psychology as a discipline is capable of acknowledging a degree of plasticity of the brain alongside modularity (Fodor, 2001, Ridley, 1999, 2003), and that instinctive modules of the brain are 'designed' to be modified by experience and environmental stimuli (Ridley, 2003). Indeed, Ridley (ibid: 280) has made the point that, 'nature versus nurture is dead. Long live nature via nurture'.

Reference was previously made to the work of Hilary Rose and Steve Rose. They are particularly harsh critics of evolutionary psychology, and have published several papers and books (later referred to here) which arguably accuse evolutionary psychologists of producing work which is reductionist, determinist and adaptationist (Curry, 2003). It could also be argued that their *Alas, Poor Darwin: Arguments against Evolutionary Psychology,* published in 2000 is perhaps their most in-depth critique. At this point it should be firmly emphasized that both Steve Rose and Hilary Rose have unassailable reputations as scholars, and that they have both contributed greatly to the literature on evolutionary psychology. It is not the intention here to suggest otherwise, merely to cover Curry's criticisms of their respective positions on evolutionary psychology. Curry (2003: 2) suggests that the collection of essays brings critics together to 'argue that evolutionary psychology is a "fashionable ideology" whose adherents are "fundamentalists" who promote "simple minded", "socially irresponsible", "culturally pernicious" explanations of human behaviour that rest on "shaky empirical evidence, flawed premises and unexamined political presuppositions"'. In what follows, we will briefly consider the main premises of evolutionary psychology and the Roses' five main arguments against it together with their account of the politics of the discipline, as outlined by Curry (ibid).

According to Dawkins (1976), modern evolutionary biology revolves around the idea that genes build organisms in order to make further copies of themselves. The actual design of organisms reflects the problems and opportunities that genes face on the path to replication. Biologists use these problems to make predictions about the forms of 'design solutions' or 'adaptations' of which organisms will be composed (Curry, ibid). It is then possible for biologists to conduct experiments to test for the presence of these adaptations (Dawkins 1986). Conversely, as Curry (ibid) suggests, 'biologists can ask whether a particular trait is an adaptation by asking whether it solves (better than chance) a problem that the organism typically faces'. These processes are sometimes referred to as 'engineering' and 'reverse engineering' (Ridley, 1999, 2003). As Bennett (1995) and Tooby (1999) have made clear, 'engineers' begin with a problem and endeavour to design something that will solve the problem. 'Reverse engineers' begin with 'a widget' (Curry, ibid) and endeavour to work out the nature of the problem. Evolutionary psychology adopts such an adaptationist approach when investigating the design of the human mind. As Curry shows, evolutionary psychologists can begin with a problem 'that would have been recurrent in the lives of our ancestors – such as how to choose fertile mates, or how to maintain co-operative alliances' (ibid). They are then able to suggest alternative solutions, and design experiments to test for them. As Tooby, Cosmides et al (1992: 10) suggest, 'If one knows what adaptive functions the human mind was designed to accomplish, one can make many educated guesses about what design features it should have, and can then design experiments to test for them'. For example, using evolutionary theory, comparative data and the ethnographic record, Symons (1979, 1995) was able to make a number of predictions about the evolved design of human sexual psychology. As Curry (ibid) shows, these predictions were put to the test and largely confirmed by a survey of 10,000 individuals from 37 different cultures conducted by Buss (1994). The process can run in reverse, according to Curry, whereby 'previously mysterious psychological devices can be illuminated by revealing the function that they are designed to perform' (ibid). An example of this is how Cosmides and Tooby (1992) identified a quirk of human psychology, discovered years earlier on the Watson Selection Task, through a series of experiments as a 'cheater-detection mechanism', a device that game theory predicts is 'necessary for certain forms of co-operation' (Curry, ibid).

As Curry (ibid) states, Steven Rose (2000: 247) appears to accept the basic premise of evolutionary psychology....

> The declared aim of evolutionary psychology is to provide explanations for the patterns of human activity and the forms of organisation of human society which take into account the fact that humans are animals and, like all other currently living organisms, are the present-day products of some four billion years of evolution. So far so good.

As Curry (ibid: 3) shows, Rose (ibid: 250) continues to expand upon the adaptationist approach….

> Because humans are as subject as any other organism to evolutionary processes, we should therefore expect to find such adaptations among our kind just as much as amongst the others that we study. Individual aspects of being human – from our body shape to our eyes and capacity for binocular vision – are clearly evolved features and fit us to the environment in which we live.

However, as Curry shows, the Roses (ibid) object to utilising such an adaptationist approach to illuminate the psychological mechanisms that, according to evolutionary psychologist underpin human *social* behaviour. This is because, according to Curry, Rose and Rose claim that 'not enough is known about the conditions under which our ancestors evolved to make claims about the problems that they faced, or to test whether or not particular features of human psychology are adaptations' (Curry, ibid). In Curry's words, the Roses ' also claim that the period of pre-history focused upon by evolutionary psychologists – the Pleistocene or 'Stone Age' – is 'the wrong one'' because 'there has been sufficient time since the end of the Pleistocene for significant evolutionary change in the design of the human mind' (Curry, ibid). In addition, the Roses appear to be arguing that 'evolutionary psychology's claims about universal features of human social psychology are contradicted by cultural and historical variability, and neglect the role of emotion in human mental life' (Curry, ibid). Finally, it is Curry's view that the Roses mistakenly use Daly and Wilson's (1998) research on step-parents to represent what they consider to be the empirical deficits of evolutionary psychology.

As we have seen, evolutionary psychologists make and test predictions about human psychological adaptations, based upon information about human evolutionary history. According to Curry, the Roses claim that not enough is known about the Stone Age to render such an approach credible and viable. What do evolutionary psychologists actually know about the Pleistocene, and how do they get access to the information? These are questions posed by Curry (ibid), and more recently by Owen (2006b). As Curry (ibid) and Owen (ibid) have suggested, the work of Tooby and De Vore (1987) shows how information about ancestral adaptive problems comes from a wide range of disciplines which include physical and paleo-anthropology, primatology and other cross-species comparisons, studies of contemporary hunter-gatherer societies, and game-theory models of social interaction. As a result of the work of those such as Cosmides and Tooby (1997), we know that our ancestors nursed, had two sexes, hunted, gathered, chose mates, used tools, had colour vision, bled when wounded, were predated upon, were subject to viral infections, were incapacitated from injuries, had deleterious necessives and so were subject to inbreeding depression if they mated with siblings, fought with each other, cooperated with each other, lived in a biotic environment with predatory cats, venomous snakes and plant toxins. They were omnivorous, ground-living primates and mammals with helpless infants, long periods of bioparental investment in offspring, and an extended period of physiologically obligatory female investment in pregnancy and lactation.

As Curry (ibid: 3-4) suggests, in order to claim that this body of accumulated knowledge is inadequate as a starting point for evolutionary psychology, 'the Roses would need to show that current research on the Pleistocene is insufficient or unreliable' or the Roses 'would need to argue that the conditions of the Pleistocene are in principle 'unknowable', and hence an evolutionary psychology will never be possible'. However, as Curry (ibid: 4) suggests, 'no such arguments are forthcoming'. According to Curry, the Roses 'do not provide any criticism of any of the actual methods' utlised to investigate the Pleistocene, 'nor do they demonstrate that any of the actual assumptions about ancestral conditions employed by evolutionary psychologists are false, or unreasonable' (Curry, ibid). As Curry (ibid) goes on to suggest, the Roses do not 'take issue with any of the predictions or discoveries about adaptations for social life' (the reference is to infanticide, concealed ovulation, sperm-competition, uncertainty of paternity, material-fetal conflict, mate-guarding, 'theory of mind') 'that evolutionary psychologists have made'. In particular, Curry (ibid) suggests that Steven Rose (2000: 253) 'merely follows Stephen Jay Gould in suggesting that accounts of ancestral conditions are little more than Just-So stories'. For Curry, Hilary Rose's (2000: 118) quip to the ends that evolutionary psychologists offer up a vision of the Stone Age 'that owes more to the Flintstones than to serious scholarship', fails to provide contrary evidence. 'The Roses' contention seems to be merely that finding out what the Pleistocene was like is difficult, and that nothing is certain. But then what branch of science is any different?' (Curry, ibid: 4). In Curry's view, the Roses are not so much arguing the point as asserting it.

Owen (2006b: 908) shows how Rose and Rose (2000) 'take issue with the idea, expressed by Nesse and Williams (1994), that there is a mismatch between 'stone-age minds' and the 'space age world'. Nesse and Williams (ibid) suggest that human evolution is an extremely slow process, and therefore evolutionary psychologists expect human psychology to be 'designed' to deal with conditions experienced during the million or so years that humans spent as hunter-gatherers, but not with the 'novel' problems encountered during the relatively brief period of roughly 10,000 years that the species has spent as farmers/postoralists, or the even briefer period spent in modern, industrialised, urbanised societies. Nesse and Williams (ibid) have claimed that the mismatch between 'stone-age minds' and 'space-age world' is part of the reason why human behaviour does not reliably result in increased reproductive success. They cite as an example, our preference for sugars, salts and fats evolved at a time when such valuable nutrients were relatively scarce, and we could not get too much of them. Conversely, these things are abundant in the modern world, and the same properties lead us to suffer from obesity, high, blood pressure and tooth decay. The Rose's (ibid) second argument takes issue with the assumptions of a 'mismatch', claiming that there *has* been enough time since the Pleistocene for some substantial evolutionary change in the 'design of the human mind'.

Rose (2000: 253-54) suggests that....

> Evolutionary modern humans appeared 100,000 years ago. Allowing 15-20 years as a generation time, there have been some 5,000-6,000 generations between human origins and modern times'. Noting that rapid evolutionary change is commonly observed in non-human species, Rose concludes that the assumption that humans have not changed significantly in the last 10,000 years, 'does not bear serious inspection.

Firstly as Curry (ibid: 4) suggests, given that 'the issue at hand is the possibility of evolutionary change during the 10,000 years since the end of the Stone Age, it is not altogether clear why Steven Rose makes an argument about the possibility of change in the last 100,000 years.' For Curry [ibid: 11], this is evidence of the Roses' 'wider confusion about the dates of the Pleistocene'. As Curry (ibid: 11) further suggests, '*In Alas, Poor Darwin* they suggest that the Pleistocene was the period between ''100-600,000 years ago'' (Rose and Rose 2000: 1). In their *New Scientist* article, the Pleistocene is ''between 10,000 and 1.6 million years ago'' (Rose and Rose 22/6/2000). And in their *Guardian* article, they suggest that there has been ''100-600,000 years *since* the Pleistocene'' (Rose and Rose 13/7/2000)'.

Secondly, Curry (ibid: 4-5) suggests that Steven Rose's argument for the speed of evolutionary change rests on 'laboratory and field experiments in species varying from fruit flies to guppies (which) give rates of change of up to 50,000 darwins... English sparrows transported to the south of the USA have lengthened their legs at a rate of around 100,000 darwins, or 5 per cent a century' (Rose, 2000: 253). A 'darwin' refers to changes in the mean of the natural log of morphological characters divided by the elapsed time in terms of millions of years over which the change has occurred (Thain and Hickman, 1995). According to Curry (ibid: 11), 'at no point does Rose explain how a measure of morphological change can be applied to the kinds of information-processing mechanisms that evolutionary psychologists study'. Curry questions whether Rose is, 'perhaps thinking of changes in brain size – but what would it mean for a software package to become 5% longer?' Even if such change had occurred, 'Rose does not explain how or why this would undermine the claim that the mind is adapted to ancestral conditions' (Curry, ibid). Rose's contention is 'presumably that if a sparrow's leg can become 5% longer over 100 years (approximately 100 generations), then it is reasonable to suppose that humans may have changed (in some unspecified way) by 25% over 500 generations' (Curry, ibid: 4-5). As Curry (ibid: 5) makes clear, for this to be possible, the period since the Pleistocene would have 'had to have been characterised by the kind of consistent, uniform, directional selection pressures that led to the changes in the sparrow's legs'. According to Curry (ibid), Rose 'does not point to any such selection pressures'. On the contrary, Rose (ibid: 253-4) appears to argue that the period since the Pleistocene has been characterised by 'very rapid changes in human environment, social organisation, technology and mode of production' [Curry, ibid].

Thirdly, as Curry (ibid: 5) suggests, quite apart from the question of how quickly evolution could conceivably occur, 'there is the question of whether any such change has *actually* occurred in the human line in the last 10,000 years'. According to Curry, Rose (ibid) presents no evidence of selection pressures. Rose [ibid: 253] concludes that, 'we really have no idea whether the 6,000 or so generations between early and modern humans is 'time enough' for substantial evolutionary change. We don't even know what 'substantial' might mean in this context'. As Curry (ibid) suggests, an obvious test of the Roses' theory would be to look for 'genetic differences in populations that have lived under different conditions since the beginning of the agricultural revolution'. If the Roses are correct, and there has been sufficient time for significant evolutionary changes in the 'design' of human psychology since the Stone Age, then, 'to the extent that different groups occupied different environments, one would expect them to have evolved in different ways' (Curry, ibid). Curry goes on to suggest that groups of humans that remained in Africa might be expected to differ from those that migrated to the Russian steppes, for example. By implication, hunter-gatherers should posses a different set of 'mental tools' from agriculturalists and industrialists. In the words of Curry (ibid), 'the Roses do not discuss this question, or present any evidence that bears on its answer'.

Crawford and Anderson (1989) have suggested that one implication of the 'evolutionary' view that the human mind took its current form in the Pleistocene is that all modern humans share a universal human nature. In answer to how evolutionary psychologists reconcile such a claim with 'the manifest diversity of behaviour and culture found around the world', (Curry (ibid) argues that psychological mechanisms are 'condition dependent', in other words, the behaviour they produce will be different under different conditions. As Curry suggests....

> Suppose for example, that human psychology operates in part according to the rule ''*If* resources are scarce, *then* adopt a more aggressive approach to acquiring them''. On this basis one might expect over levels of aggression to vary according to the current economic or ecological circumstances of the people under consideration. Evolutionary psychologists look at, amongst other things, permutations in behaviour in order to work out what the underlying rules are and how they operate. This research, which necessarily involves cross-cultural studies, commits evolutionary psychologists to a strongly ''environmentalist'' position: the idea that differences in behaviour are largely the product of differences in environmental-physical, social or cultural factors.

Perhaps here Curry echoes Ridley's (1999: 53) point (in the 'Building Bridges' section of this Chapter) to the ends that 'genes need to be switched on', and 'external events – or free-willed behaviour – can switch on genes'. In other words, we should strive to avoid what Owen (2006a, 2006b) and Powell and Owen (2005) call *genetic fatalism* (and which is incorporated into the metatheoretical framework employed here) – the tendency to associate genetic predisposition with inevitability. This point aside, the Roses' third argument against evolutionary perspectives appears to be that cultural and historical variability in cultural forms *refutes* the view from evolutionary psychology that all humans share a universal species-typical psychology. Steven Rose (2000: 262-3), referring to historically-recent changes in female mate-preferences, levels of violence, and fertility in several hunter-gatherer populations, remarks 'Each of these societies has undergone rapid economic, technological and social change in the last decade. What has happened to the evolutionary psychology predictions? Why have these assumed human universals suddenly failed to operate?' However, as Tooby and Cosmides (1992: 46) suggest, 'The recognition that a universal evolved psychology will produce variable manifest behaviour given different environmental conditions exposes this argument as a non-sequitur'. As Curry (ibid: 6) makes clear, in order to make their point effectively, the Roses, 'would need to show not only that behaviour changes, but that behaviour changes in ways that are *not* predicted by the evolutionary psychologists' account of the mechanisms responsible', and 'even if the Roses did present such evidence, they would be refuting a particular hypothesis about a particular mechanism, not the entire evolutionary approach to psychology' (ibid: 11). It is Curry's view (ibid: 6) that 'by presenting social and cultural explanations as *alternatives to* - rather than *continuous with* - biological explanations, the Roses perpetuate the myth that genes can operate only in a rigidly 'deterministic' fashion'.

Steven Rose's fourth argument appears to be along the lines that evolutionary psychology 'neglects the role that emotion plays in human mental life' (Curry, ibid). Rose (2000: 262) claims that….

> Emotional mechanisms and indeed their expression are evolved properties, and several neuroscientists have devoted considerable attention to the mechanisms and survival advantages of emotion. So it is therefore all the more surprising to find this conspicuous gap in the concerns of evolutionary psychologists.

As Curry (ibid: 6) shows, the entry for 'Emotion' in the index of the evolutionary psychologist Steven Pinker's work *How the Mind Works* – to which both the Roses refer – reads as follows:

Emotion, 65, 143, 315, 363-424; adaptive function, 143, 370-374; facial expressions, 273, 365-366, 374, 379, 414-416, 546; hydraulic model, 57, 65, 551; and imagery, 285; in music, 529, 531-532, 533-534; neuroanatomy, 371-372; phylogeny, 370-371; universality, 364-369. See also Anger, Beauty; Disgust; Exhilaration; Fear; Gratitude; Grief; Guilt; Happiness; Honor; Jealousy; Liking; Love; Passion; Self-control; Sexual desire; Shame; Sympathy; Trust; Vengeance (Pinker, 1997: 637).

According to Curry, the Roses also 'neglect to mention' Tooby and Cosmides' discussion of emotion; Buss's work (1994, 2000), Nesse's (1990) paper on the 'evolutionary explanations of emotions'; Ekman's work on the facial expression of emotion, including his recent work involving an edited version of Darwin's *The Expression of the Emotions in Man and Animals* and others. In Curry's words, 'Steven Rose appears to have forgotten that he has debated the evolution of human emotions with Steven Pinker' in Pinker and Rose (21/1/1998). Yet, in Rose and Rose (13/7/2000), the Roses appear to criticise the claim from evolutionary psychology that there is a universality to human emotions. Curry (ibid: 6-7), argues that Steven Rose's claim that evolutionary psychology 'neglects the emotions' rests on an, 'elementary misunderstanding of what it means to be an "information processor"'. For example, Rose (2000: 261) suggests that….

> It is not adequate to reduce the mind/brain to nothing more than a cognitive 'architectural' information-processing machine. Brains/minds do not just deal with information. They deal with living meaning… The key here is emotion, for the key feature which distinguishes brains/minds from computers is their/our capacity to experience emotion.

However, as Curry (ibid: 7) suggests, the adoption of an 'information processing' or 'computational' approach to the mind 'does not commit you to a desktop-computer view of psychology, any more than charting the parabola of a falling apple commits you to inferring that the apple is performing calculus'. As Churchland and Grush (1999) have made clear, 'Information theory' is merely a branch of mathematics utilised in order to capture how a particular system's inputs map onto its outputs. The theory does not imply that the most accurate explanation of human brain function will actually be in computational terms. As the authors have suggested, in an abstract sense, livers, stomachs and brains *all* 'compute', so consequently 'information' need not exclude 'emotion'. Further evidence (cited in Curry (ibid: 7) from Tooby and Cosmides (1992: 65) can be provided on this point….

> It is important to keep in mind exactly what we mean by the cognitive or information-processing level... some researchers use it in a narrow sense, to refer to so-called 'higher mental' processes, such as 'emotion' or 'motivation'... In contrast... we use terms such as cognitive and information-processing to refer to a language or level of analysis that can be used to precisely describe any psychological process: reasoning, emotion, motivation, and motor control can all be described in cognitive terms.

Curry (ibid) makes the point that *Alas Poor Darwin* does not offer, in the form of research, evidence to 'refute any of the empirical claims of evolutionary psychology', rather 'it carries little sustained discussion of any empirical work'. This is, of course, Curry's own personal opinion. However, Curry does acknowledge that there is an exception here in Hilary Rose's arguments against the work of Daly and Wilson (1998). In the latter work, the authors summarise findings that children are at significantly higher risk of sexual/physical abuse and murder from step-parents than from genetic parents. Daly and Wilson's (ibid) explication is that parental investment is a costly resource, and that Darwinian natural selection has played a part in the 'design' of parental psychology to preferentially 'invest' in one's genetic children, and to be reluctant to 'invest' in children that are not one's own genetic offspring. As Curry (ibid) shows, Daly and Wilson (ibid) make the point that step-relationships can be the site where such reluctance may manifest itself in less than harmonious relations. In extreme cases, such neglect may result in severe abuse or even murder. Although Hilary Rose (2000: 122) does not dispute this higher incidence of abuse and murder by step-parents, her contention is that, 'rather obvious matters of context' pertaining to 'psychological strain' and 'financial pressures' of starting second families, 'explain better why some men ill-treat their partner's children'. Steven Rose (2000: 260) criticises Daly and Wilson (ibid) for 'ignoring (these) much more obvious proximal causal processes'. Also writing in *New Scientist* (22/6/2000), the Roses claim that their 'obvious' alternative causes 'are better grounded than untestable evolutionary speculations' [cited in Curry, ibid:7-8]. Hilary Rose, according to Curry (ibid: 8), 'does not elaborate on these "obvious" alternatives, nor point to research that has "grounded" them'. According to Curry, Hilary Rose 'does not make any references to any other studies of child abuse whatsoever'. Curry (ibid) suggests that Rose appears, 'entirely unaware that these "obvious" explanations have been put to the test, and that they failed'. Some evidence for this claim may lie in Daly and Wilson's (1998: 29) work on the effects of poverty in which, 'this initially plausible hypothesis was rejected, for it turned out that the distribution of family incomes in step-parent homes in the United States was virtually identical to that in two-genetic-parent homes'. Daly and Wilson (ibid) have also examined the effects of numerous other confounding variables such as 'the age of the child, age of the mother, whether the step-father was present at the birth, and the personalities of people who remarry' (Curry, ibid: 12). Daly and Wilson (1998: 45-46) also discuss the apparent 'anomaly of adoption'. Daly and Wilson conclude....

> All in all, although several additional risk factors were identified, step-parenthood held its place as the most important predictor (of serious child abuse), and its influence was scarcely diminished when the statistical impacts of all the other risk factors were controlled (p.31).

As Curry (ibid: 8) shows, Daly and Wilson (1988) in their major work, *Homicide*, also discuss how 'umbrella' or 'blanket' explanations (overarching theories) such as 'strain' do not explain such nuances as the fact that in families where there are both step-children and genetic-children present, it is the step-children who are invariably singled out for abuse and even murder. According to Curry, Hilary Rose (ibid) has complained that Daly and Wilson's (2001) data do not distinguish between a step-child whose parent has died and a step-child whose parent is absent because of divorce, or between married step-parents and 'casual lovers'. However, as Curry (ibid) suggests, Rose, 'does not suggest *any* reason why, or any evidence *that*, the distinctions she suggests – or indeed, any of the infinite number of distinctions that one could make – might matter'. Hilary Rose (2000: 121) claims that Daly and Wilson's work receives 'no support from the primatologist Sarah Hrdy, whose work on langur monkey infanticide is key to their thesis'. Hrdy was quoted in *New Scientist* as saying, 'Human violence towards babies and infants may be tragic but it's nothing like what a langur male is doing'. As Curry (ibid) points out readers of *Cinderella* might be surprised to discover that Daly and Wilson have the same perspective on how Hrdy's thesis relates to their own evolutionary position. As Daly and Wilson (1998: 37-38) have said....

> Human beings are not like langurs or lions. We know that 'sexually selected infanticide' is not a human adaptation because men, unlike male langurs and lions, do not routinely, efficiently dispose of their predecessors' young... Quite unlike the situation in langurs or lions, human stepfamilies exist in all societies, and most stepchildren survive them.

Curry (ibid) has suggested that none of the Roses' 'arguments against evolutionary psychology' successfully establishes that the discipline is based upon and relies upon spurious empirical evidence or premises. However, next we need to consider the Roses' accusation that evolutionary psychology relies upon 'unexamined political presuppositions' (Curry, ibid). Steven Rose (1997) and Rose, Lewontin et al. 1984) appear to promote the view that there is 'more to' evolutionary explications of human behaviour than mere science. Rose and Lewontin et al (ibid) offer an arguably Marxist critique of evolutionary biology; and Rose himself (ibid: pix) warns that...

> The rise of the present enthusiasms for biologically determinist accounts of the human condition date back to the 1960s. They were not initiated by any specific advance in biological science, or powerful new theory, but harked back instead to an earlier tradition of eugenic thinking which.... had been eclipsed and driven into intellectual and political disrepute in the aftermath of the war against Nazi Germany and its racially inspired holocaust.

According to Curry (ibid: 9), *Alas Poor Darwin*, 'marks a complete reversal from this earlier position', with Hilary Rose now conceding that evolutionary psychology, 'eschews any notion of race' and that it is compatible with a reasonably broad range of political viewpoints including, 'Peter Singer's Darwinian Left, Matt Ridley's free marketeering, Helena Cronin's feminism, Francis Fukayama's call for state intervention to tackle unemployment' etc. As Curry (ibid) makes clear, each of these scholars/researchers provide evidence that, facts and values can be kept separate; that one's political goals do not dictate one's science (or vice versa), but, as Trivers (1981) has suggested, once political and social objectives are agreed upon, science can help one achieve these goals. Curry (ibid) suggests that, in order to demonstrate that evolutionary psychology is politically-motivated, the Roses must prove that 'the political views of a researcher have led to a scientific error', but as yet, in his view, 'no such demonstrations are forthcoming' in their work.

Curry (ibid: 9) makes the powerful point that the Roses fail to show that evolutionary psychology rests upon 'shaky empirical evidence' or 'flawed premises'. Curry's work suggests that....

> Evolutionary psychologists use accounts of pre-history corroborated by numerous diverse disciplines; their assumptions about the rate of evolutionary change are in keeping with standard assumptions in evolutionary biology; their claims about human universals are consistent with, and go some way towards explaining, cultural and historical variability; emotions have been the subject of a great deal of evolutionary research; and Martin Daly and Margo Wilson's findings have withstood over two decades of testing and scrutiny. Nor do the Roses provide any reason for thinking that evolutionary psychology is a "fashionable ideology" motivated by "unexamined political presuppositions".

As previously stated, our task here is to establish whether there is sufficient evidence from evolutionary psychology to justify modifying Sibeon's anti-reductionist, anti-relativistic, meta-theoretical framework so that it entails focusing upon the *biological variable* (the evidence that human behaviour is, at least in part, genetically-determined). The work of Owen (2006a, 2006b, 2007a] has made the case, based in part upon the work of Ridley (1999, 2003), that we can talk in terms of human instincts, and that genes can be switched on, and external events – or free-willed behaviour – can switch on genes. Ridley (2003) makes the case for abandoning the 'nature versus nurture' debate and instead champions the alternative conception of 'nature via nurture'. He believes human behaviour 'has to be explained by both nature and nurture' (ibid: 3). The author goes on to suggest that the discovery of the human genome has changed everything, 'not by closing the argument or winning the battle for one side or the other, but by enriching it from both ends till they meet in the middle' (ibid). In other words....

> Genes are designed to take their cues from nurture. To appreciate what has happened, you will have to abandon cherished notions and open your mind. You will have to enter a world where your genes are not puppet masters pulling the strings of your behaviour; a world where instinct is not the opposite of learning, where environmental influences are sometimes less reversible than genetic ones, and where nature is designed for nurture (Ridley 2003: 4).

Later in the Chapter we incorporate *the biological variable* into the metatheoretical framework, having considered further evidence from Ridley (1999, 2003) and others which suggests that human beings can be simultaneously free-willed (or possessing *agency*) and motivated by instinct and culture. The Chapter, having drawn upon further evidence that nurture depends upon genes too, and how genes need nurture, plus the relationship between 'free will', genes and 'freedom', concludes with the intention of applying some of the terms to the study of selected examples of human biotechnology in the final Chapter. However, at this point we need to re-examine some of the issues around the *dualism/duality* debate. The emphasis here is on whether we need a dualistic conception of 'biology' and 'society' or whether we should adopt an approach which favours a *duality* of 'biology' and 'society' alongside 'agency/structure' etc, in order to contribute towards a study of biotechnology.

DUALITY VERSUS DUALISM REVISITED

As observed previously, Williams (2003: 556) has recently stated that 'material-corporeal' sociologists engaged in the project of 'building bridges' between sociology and biology appear to be united in the desire, 'to challenge dualism on all fronts'. Benton (2003: 292) has also called for an approach which would 'dissolve' dualistic oppositions ('nature' and 'society'). One reflects that this is not the same as Ridley's (1999, 2003) approach which favours a conception of *'nature via nurture* Ridley argues that human instincts do exist, but that genes can be 'switched on' by environmental stimuli. Writers of the 'embodied school' such as Newton (2003: 29) appear to demonstrate both an *oversocialised* approach and *genetic fatalism* in their criticisms of approaches which acknowledge the existence of instincts in human beings, and even partial biological causality. Newton's criticisms of Sahlins (1972) and Pollock (1988) centre around the authors' portrayal of stress as the result of 'natural' instincts. Newton (ibid) suggests that, 'such arguments present a *crude dualism* that reduces the 'problem' of present day social complexity to the outmoded biological body', and goes on to argue that 'the biological reductionism of such discourse' is closer to 'reductionist sociobiology than it is to, 'a putative non-reductionistic 'biological sociology''. Considering the evidence presented so far, and the evidence which is provided later in the Chapter from Ridley (ibid), it is hard to envisage a 'biological sociology' which denies the existence of instincts. Writers of the 'embodied' school such as Williams (ibid), Benton (ibid), Shilling (1993) and Freund (2001) construct *oversocialised* conceptions of 'the body' in the sense that they appear to reject unlearned, genetically-based behaviour outright, and tend to view the idea of genetic predisposition as 'inevitability', hence Newton's references to 'biological reductionism' (ibid). The 'embodied' school appear to argue for a *duality of structure* in which 'biology' and 'society' are seen as 'two sides of the same coin'. For example Shilling's analysis of the body as *simultaneously* biological and social. Newton (ibid: 35) sees, 'no epistemological reason to erect a barrier between the biological and the social'. Again, this is quite different to Ridley's (ibid) position because Newton rejects the idea of instincts in human beings, as we may observe in his previously noted criticisms of Sahlins's (ibid) and Pollock (ibid). Ridley's approach is to acknowledge the existence of instincts but to acknowledge that genes are 'designed to take their cues from nurture' (2003: 4). Newton's and Shilling's (ibid) positions appear to be similar to Giddens's (1991; 1982; 1984; 1991b; 1993) theory of *structuration* in the sense that they are 'elisionist' (Archer 1995: 93), collapsing 'biology' and 'society' together as Giddensian 'flat' social ontology collapses the 'micro-macro' distinction.

As Sibeon (2004: 48) makes clear, 'the contemporary debates of 'dualism versus duality'' centre around critique of Giddensian structuration theory. Giddens, in company with some post-structuralist and postmodern theorists considers the micro-macro distinction to be an 'unhelpful dualism' because it serves to polarise social scientists into opposing camps; perpetuates a tendency to associate micro with agency, and macro with constraints, in the form of social structure upon agency (1984; 139; 1993: 7). According to Sibeon (ibid), Giddens also reject micro-macro dualism because it tends to 'emphasise the difference between small groups and larger social phenomena'. Giddens appears to favour a '*duality* – not a *dualism* – of action and structure' (ibid). Structuration theory revolves around the idea of 'duality of structure', in which agency and structure are not separate 'domains'. The distinction between these elements is 'dissolved', as the intention is, 'to avoid giving primacy to either action *or* to structure and this is expressed in his concept of 'social practices' (ibid: 75). With regard to the approaches of Shilling (ibid) and Newton (ibid), in relation to their conflation of 'society' and 'biology'; these authors are effectively giving 'primacy' to environmental factors in their denial of instincts. Regarding 'social practices'; as Giddens (1984: 2) himself suggests, 'The basic domain of study of the social sciences is neither the experience of the individual actor, nor the existence of any form of social totality, but *social practices* ordered across time and space'. These 'practices' are, as Sibeon (ibid) suggests, 'a combination of action and structure' and 'structure' is both the medium and the largely unintended outcome of social action/social practices. Perhaps Sibeon (ibid) is correct when he suggests that, 'in social analysis, agency-structure and micro-macro should be employed as *dualisms* that refer to distinct, relatively autonomous phenomena'. Arguably, 'biology' and 'society' should, and can be, employed in analysis as distinct phenomena, although it is important to acknowledge how human social actors can be both free-willed and motivated by instinct *and* culture. To reiterate, it does not seem feasible to deny instinct altogether in the *oversocialised* fashion of Newton's (ibid) work. As Ridley (2003) suggests, genes predetermine the broad structure of the brain, but also absorb formative experiences, react to social cues and even run memory. Nevertheless, although genes and 'environment' both play a part in the explanation of human behaviour, it may sometimes be necessary to analyse them separately. Adopting a 'duality of structure' viewpoint makes this impossible to achieve. For example, Shilling's (ibid) analysis of the human body as simultaneously 'biological' and 'social' serves to collapse the terms into an amalgamated 'whole' within which elements cannot be separated. Sibeon (ibid) makes the cogent point that Giddens 'makes it impossible' to investigate the elements of micro, macro, agency and structure, separately. Perhaps the same can be said about Shilling and other 'material corporeal' sociologists such as Benton (ibid), Williams (ibid) and Freund (ibid). One reflects that it is desirable to be able to investigate biology and the social in terms of their mutual influence upon each other over time, or as Sibeon (ibid) has said of agency, structure, micro and macro, 'to ascertain the relative impact of each upon any given social situation'. In our case, the acknowledgement of the *biological variable* will play a part in the application of the metatheoretical framework to a study of biotechnology.

Margaret Archer (1988; 1995; 1998) plays a central part in the contemporary 'duality versus dualism' debates. She rejects Giddensian concepts of the duality of structure, arguing (as we saw in Chapter Two) for a dualism of structure/action. Here we have a *temporal* separation between structure and action: 'structure precedes the action which reproduces or elaborates it' (Sibeon ibid: 102). For Archer (1995: 01), all social analysis and theorising must be based upon dualism as duality is regarded as *'central conflation'*. Benton's (2003: 292) recent attempts to, 'dissolve the dualistic opposition between 'society' and 'nature'' appear to be examples of what Archer regards as *'central conflation'*.

Conversely, Mouzelis has argued (1995: 196) that there *is* room for duality in social theory and analysis, stating that, 'I shall continue to use the term "structure" in the way Giddens has defined it'. However, as Sibeon (ibid) has pointed out, Mouzelis does not so much reject structuration as 'reconstruct it' so that it entails including both dualism *and* duality. Mouzelis has defined duality as a concept that refers to situations, 'where actors have no cognitive distance from 'rules' (he has in mind situations where there is an unreflective, routine drawing upon rules)', and 'dualism for Mouzelis is a term that indicates, again in a paradigmatic sense, scenarios where actors draw reflexively upon rules for theoretical-analytical, strategic or monitoring purposes' (Sibeon ibid: 103). Giddens (1993: 6) maintains that Mouzelis misunderstands his concept of the duality of structure, and that no useful purpose can be served by equating reflexivity with dualism. As Sibeon (ibid) makes clear, Archer conceptualises dualism as a 'temporal separation of structure and agency, and duality as a collapsing of this separation'. For Mouzelis (and under Mouzelis's meta-theoretical framework, Giddens too) the distinction between dualism and duality refers to 'differences in actors' orientations – reflexive or otherwise – to structure' (Sibeon ibid). Arguably, the framework employed here in *Social Theory and Human Biotechnology* [a metatheoretical framework concerned with methodological generalisations as opposed to substantive generalizations] has more in common with the work of Sibeon and Archer in terms of favouring *dualism* and the 'micro/macro', 'biology-society' distinctions than it has with the work of Mouzelis or Giddens. However, as Sibeon (ibid) points out, Mouzelis's position raises some important points.

Sibeon's interpretation of Mouzelis's position is along the following lines. Mouzelis (1995: 118) claims that Giddensian concepts of 'structure' refer to sets of rules (in a 'virtual' sense') that exist outside time/space. Sibeon goes on to suggest that Mouzelis considers that Giddensian concepts of 'systems' refer to 'syntagmatic, 'actual' patterned relations in time-space'. Sibeon's (ibid) interpretation of Mouzelis suggests that it is, 'via the processes of structuration that the paradigmatic becomes syntagmatic': and, in Mouzelis's (1995: 118) words, 'the term structuration signifies the process by which structures lead to the constitution of social systems'. Mouzelis (ibid: 120) has suggested that social actors relate to 'structure as 'virtual rules'', in terms of duality *and* dualism with 'both equally indispensable'. Here, as Sibeon (ibid: 104) suggests Mouzelis is not rejecting Giddensian duality of structure, rather he 'extends and reconstructs' the concept. The central point is that Mouzelis sees structuation theory as 'incomplete' (Sibeon ibid) as the idea of duality is employed to the exclusion of dualism. Mouzelis's (1991: 100) critique of structuration contains four theoretical terms which are 'continua rather than either-or states of affairs', and they refer to, 'variation in the relation of actors to 'virtual' structures (as 'rules') (the paradigmatic aspect) and, syntagmatically, to 'actual' social objects or systems' (Sibeon ibid). The concept of paradigmatic duality concerns duality of structure, where social actors are seen as, 'drawing on more or less "hidden" rules in a natural-performative manner' (Mouzelis 1995: 121); and 'it is on this partial case (that is, paradigmatic duality) that Giddens bases his duality-of-structure notion' (ibid: 121). *Paradigmatic dualism* is, according to Sibeon (ibid), 'where actors, for theoretical-analytical, strategic, or monitoring purposes distance themselves ('stand back') reflectively from a 'virtual' body of rules (that is, in Giddens's and Mouzelis's terms, from structure)'. What is of great importance here is the distance between rules and social actors. *Syntagmatic duality* is, 'where the actor is vital to the existence of a social context (that is, to the existence of an 'actual' – not virtual – social system, social context or social game (s))' (Sibeon ibid: 104-105). Here, the social actor plays a part in making/reproducing interaction in a social system. *Syntagmatic dualism*, 'refers to states of affairs where the actor is not vital to, or has little effect upon the social content (for example, an office cleaner or junior clerk in a multinational corporation)' (Sibeon ibid: 105). As Mouzelis (1991: 100) himself has made clear, syntagmatic dualism can be said to apply in situations where, 'actors' orientations focus on interactions or social systems to the production or reproduction of which they contribute but slightly'. Arguably, as Sibeon (ibid) suggests, despite the 'sophistication' of Mouzelis's typology, the notion of duality as expressed here is actually flawed. It is perhaps more fruitful to endorse Archer's concept of *central conflation*, and employ the micro-macro distinction in sociological analysis. In the case of the synthetic, meta-theoretical framework developed here and later applied to the study of selected examples of human biotechnology, perhaps one should also maintain the 'biology-society' distinction as well as the dualisms of agency- structure and micro-macro.

As we saw in Chapter Two, Derek Layder's (1997) work has contributed greatly both to debates around metatheory in a post-postmodern landscape, and to debates pertaining to the 'duality versus dualism' question. Layder's (ibid) cogent criticism of Giddensian rejection of the agency-structure distinction suggests that, 'Giddens cannot examine how far, in concrete situations, actors are influenced or constrained by social contexts (Sibeon ibid: 106). If one were to employ such a concept of ontological duality in an attempt to 'build bridges' between 'the social' and biology, one would be melding the elements so tightly together that it would be impossible to investigate them separately. Benton's (2003: 293) call for a, 'multi-faceted, multi-dimensional viewpoint' which 'avoids the simplistic opposition of 'nature' versus 'nurture' in relation to health' sounds superficially akin to an ontologically-flexible, meta-theoretical approach. However, Benton's concept of the, 'complex intertwining' of actors' 'bodily contribution, including their genetic endowment, the sociological relations characteristic of their society, and their own position within that society' appears to be a 'sophisticated' case of *conflation*. As Sibeon (ibid: 113) makes clear, 'Layder's commitment to dualism relates to his contention that social reality is best described in terms of a stratified or 'depth' social ontology that involves rejection of any attempts to collapse distinction between micro and macro (or between agency and structure)'. Layder's social ontology, as we have seen in Chapter Two, includes four 'domains', and each has 'relative autonomy' (Sibeon, ibid). The terms *psychobiography* and situated *activity* refer to the subjective dimensions of the social context. The terms *social settings* and *contextual resources* refer to the objective dimension. *Psychobiography* is a particularly useful term and concept for those who would make links between biology and sociology, as it refers to unique, asocial components of the person. Here, Layder appears to recognise the existence of, at least in part, biological variables and biological causality in human behaviour. This, however, is an observation made with caution. Social *settings* are locales within which situated activity occurs, and the term *contextual resources* describes macro-distributions and ownership of resources, together with widespread discourses and practices. The four 'domains' or layers have 'vertical' or 'lateral' dimensions, and micro, mezo and macro levels of social process are acknowledged. In sum, there are definite links in terms of *dualism* between the framework employed here and the work of Sibeon, Layder and Archer in particular. To reiterate, for the development of the framework it is desirable that we avoid a Giddensian *duality of structure* which would serve to collapse the distinction between agency and structure, micro and macro, and in this context, biology and the social. Rather in social analysis, particularly in the case here of a later application of the framework to the study of human biotechnology, agency-structure, micro-macro, and biology/'the social' should be employed as *dualisms* that refer to discrete phenomena. This does not mean, of course, that connections cannot be made between variables. In the case of biology and 'the social', it is the case that genes can be 'switched on' by environmental stimuli (Ridley, 1999; 2003).

INCORPORATING THE BIOLOGICAL VARIABLE

In what follows we examine further evidence pertaining to the relationship between free will (agency), genes and 'freedom' from the work of Ridley and Hamer and Copeland (1999), in particular. The case is made for incorporating the *biological variable*, as a theoretical concept, into the metatheoretical framework alongside *agency-structure, micro macro* and *time space.* We have already considered the evidence from evolutionary psychology to the ends that genes *do* play a part in human behaviour, despite the *oversocialised* resistance to the idea from the work of those such as Newton (2003) and Giddens (1993). Here, we also refer back to theorists of the 'embodied' school in relation to their attempts to 'bridge the divide' between the social and the life sciences, and we briefly apply a few terms such as *genetic fatalism* to certain theoretical approaches. It must be emphasised that the detailed application of the framework's constructs to the study of human biotechnology takes place in Chapter Five, and to a lesser extent amid discussion of general issues pertaining to human biotechnology in Chapter Four. As stated in the 'codification' section in the earlier part of this Chapter, the framework is intended to 'prepare the ground' for the study of human biotechnology, and in order to apply concepts to selected examples, we must first justify its existence, in particular, its modification (from Sibeon's original *anti-reductionist* framework) to include focusing upon genetic or partially genetic (in Ridley's (ibid) sense of 'nature via nurture') causality.

To recap, the metatheoretical framework employed here focuses upon the original (in Sibeon's work) concepts of *agency-structure, micro-macro, time-space* and Foucauldian *power*, and we are considering whether to incorporate the *biological variable*; the evidence from evolutionary psychology and behavioural science for a role for biological causality, at least in part, in explanations of human behaviour. The framework is an attempt at metatheory, which, in the spirit of Sibeon's original *anti-reductionism* seeks to avoid *relativism* and favours methodological generalisations as opposed to substantive generalisations, and *dualism* rather than *duality of structure*. It arises out of a critique of the original forms of illegitimate theoretical reasoning identified by Sibeon; the 'four cardinal sins' of *reductionism, reification, essentialism* and *functional teleology*, alongside 'new' terms such as *genetic fatalism*, or the tendency within mainstream social science to equate genetic predisposition with inevitability and, the *oversocialised gaze*, which refers to harshly 'environmentalist' accounts which deny biological causality, instincts etc altogether. The framework has been employed in the work of Owen (2006a, 2006b) and Powell and Owen (2005) to examine the possibilities of constructing a post-Foucauldian sociology of social gerontology, to study selected examples of biotechnology, and to critique the 'Bio-medical' model of ageing, respectively. The task of this particular metatheoretical framework is to hopefully generate conceptual devices that will inform the development of substantive theories pertaining to human biotechnology. Arguably, the 'sensitising device' is nearer in spirit to the work of Dickens (2000) in the sense that it seeks to 'embed' human beings within their 'natural' environment whilst acknowledging both genetic and environmental causality, rather than the work of the 'embodied' school (Benton, Freund, Williams etc). The synthetic framework seeks to address the 'reality' that genes *do* influence human behaviour, whilst acknowledging that 'environment' is still extremely important.

As previously stated, the framework arises out of a critique of several illicit forms of theoretical reasoning, one of which is *genetic fatalism*. This tendency to equate biological causality with inevitability is fairly widespread in the contemporary literature pertaining to genomics and the possibility of a 'marriage' between biology and the social. Newton's (2003: 35) recent, cautious 'call' for a *biological sociology* contains several examples for which evidence is shortly provided. Newton acknowledges Shilling's (1993) 'starting point' for a marriage between the respective fields (an analysis of the human body as simultaneously biological and social), but also illustrates methodological and epistemological problems which hamper any attempts to move beyond Shilling's position. Newton (ibid: 36) warns against reliance upon, 'psycho-physiological research in order to enter the biological terrain', and draws attention to 'difficulties in knowing the body or the biological'. The author makes it clear that he is not suggesting we have arrived at, 'some 'third way' between critical realism/naturalism and constructionism that can be applied reliably across both the natural and the social arena' (ibid). The main question for Newton is whether, when crossing the 'divide', we remain on the same ontological and epistemological ground. Newton (ibid: 37) makes a cogent point when he calls for 'a more sophisticated understanding of psychology and biology' because 'without that, we can neither deconstruct it, nor reconstruct it'. Williams (2003: 559) in reply called Newton's work, 'a *welcome, timely* and *topical* contribution' to evolving debates. However, there appears to be evidence for a kind of *genetic fatalism* in Newton's (ibid: 29) criticism of Benton's (1991: 17) treatment of the work of Walter Cannon (1914). Benton is of the view that Cannon's work reflects a wish to think 'holistically' about life forms. In reply, Newton (ibid) disagrees, suggesting that Cannon's arguments are based on 'a naturalistic, reductionistic and desocialised account which explains human behaviour chiefly in terms of instincts'. Cannon's (ibid: 264) suggestion that, 'the emotion of fear is associated with the *instinct* for flight, and the emotion of anger or rage with the *instinct* for fighting or attack' is arguably less of a 'reductionistic' and 'desocialised' account than Newton appears to think. Surely, Cannon is arguing for what Ridley (2003: 4) calls 'nature-via-nurture' a scenario in which genes react to social cues, rather than a form of inevitable, fatalistic determinism? Cannon appears to be suggesting that genes 'take their cue' from nurture.

As Ridley (ibid: 6) has said ....

> Genes are not puppet masters, nor blueprints. Nor are they just the carriers of heredity. They are active during life; they switch each other on and off; they respond to the environment. They may direct the construction of the body and brain in the womb, but then they set about dismantling and rebuilding what they have made almost at once – in response to experience. They are both cause and consequence of our actions. Somehow the adherents of the 'nurture' side of the argument have scared themselves silly at the power and inevitability of genes, and missed the greatest lesson of all: the genes are on their side.

Newton (ibid: 20), when referring to writers of the 'embodied' school (Benton, Freund, Williams etc), makes the point that though they eschew reductionist, naturalist and dualist arguments, they 'nevertheless draw on studies that have some or all of these characteristics'. Newton does not appear to appreciate the possibility that one can have a viable framework which emphasises *dualism* (of biology and the social) and acknowledges *the biological variable* (the evidence that some behaviour is, at least, in part, rooted in genetics) without being reductionist. As we have seen earlier, it is possible to acknowledge how genes and environment *both* play a part in explanations of human behaviour, without reducing the explanation to a single, unitary cause, and also maintain a *dualism* (of genes and environment) for purposes of analysis, thus avoiding *conflation*. Newton questions the work of Freund (1998; 854-5; 1990: 464; 1998: 277) and Williams (1998: 129; 2001: 71) for citing work on primates like baboons in the sense that, 'reference to animal studies incites a reductionistic account of human health and appropriates rather inappropriate analytic concepts' (Newton ibid: 28). Aside from the *genetic fatalism* implied by the use of the term 'reductionistic' in relation to primate studies, Newton ignores the 'scientific reality' that Homo Sapiens as a species is an evolved African ape (Dawkins, 1992).

Newton is, arguably, not alone in equating genetic determinism with fatalism in contemporary debates. Williams (1996: 41) criticises medical sociologists for, 'their tendency to relegate the body to biology'. Instead, 'sociology should be fundamentally embodied'. Aside from rejecting *dualism*, Williams appears to equate biological determinism with fatalism, hence the use of the phrase, 'relegate'. In later work, Williams and Bendelow (1998: 144) make the same mistake of associating any explanation of human behaviour which does not explicitly mention 'environment' with inevitability, when they refer to the work of Freund as 'pointing the way towards', 'a "socialised" biology rather than a reductionist sociobiology'. As one is hopefully making clear, social science appears to be dominated by those who reject biological causality, even in part. As Ridley (1999: 91-92) argues ....

> To think otherwise, to believe in innate human behaviour, is to fall into the trap of determinism, and to condemn individual people to a heartless fate written in their genes before they were born. No matter that the social sciences set about reinventing much more alarming forms of determinism to take the place of the genetic form: the parental determinism of Freud; the socio-economic determinism of Marx; the political determinism of Franz Boas and Margaret Mead; the stimulus-response determinism of John Watson and B.F. Skinner; the linguistic determinism of Edward Sapir and Benjamin Wharf. In one of the great diversions of all time, for nearly a century social scientists managed to persuade thinkers of many kinds that biological causality was determinism while environmental causality preserved free will; and that animals had instincts, but human beings did not.

Another illicit form of reasoning which, it is argued here, we should avoid when engaging in analysis is the *oversocialised* gaze. If *genetic fatalism* is the equation of biological determinism with inevitability, *the oversocialised gaze* refers to strongly 'environmentalist' accounts which seek to deny *the biological variable*, (biological causality, instincts) altogether. Surprisingly, in the age of the Human Genome Project and rapid advances in the field of genetics, many such accounts may be found within social science. For example, as the work of Owen (2006b: 907) shows, 'Giddens's (1993: 57) suggests that, 'human beings have no *instincts* in the sense of complex patterns of unlearned behaviour''. As Owen (2006a: 190) suggests, 'Foucault's (1980b) argument that sexuality is a socio-cultural creation, that sexuality as we know it is the production of a particular set of historical circumstances and obtains only within the terms of a discourse developed since the seventeenth century, is perhaps another example of the *oversocialised gaze*'. Owen (ibid) goes on to argue that Foucault's position is very similar to that of the symbolic interactionist writers, Gagnon and Simon (1973). The latter authors adopted a 'radical form of social construction theory which is extremely *oversocialised,* in arguing that there is no 'natural' sexual drive in human biological make-up' (Owen, ibid). Sexual drive must, under their terms, be regarded as a cultural and historical construction. As far as Gagnon and Simon were concerned, not only do we *learn* what 'sex' means, and what is sexually arousing to us, but we also *learn* to want sex. The authors acknowledge that the human body has a repertoire of 'gratifications' (including the capacity to have an orgasm), but this does not mean that we automatically, instinctively want to engage in them. Certain 'gratifications' will be selected as 'sexual' through the *learning* of 'sexual scripts'. From Gagnon and Simon's perspective, socialisation is not concerned with controlling innate sexual desire so that it is expressed in 'civilised', acceptable ways, but the *learning* of complicated sexual scripts which serve to specify circumstances which elicit sexual desire. From this standpoint similar to Foucauldian conceptualisations of sexuality, 'sexual desire' is a learnt social goal. Contradictory evidence, as Owen (ibid) shows, can be found in the work of Hamer and Copeland (1999: 163) who have shown how genes influence our sexual desire, how often we have sex etc and 'help make us receptive to the social interactions and signs of mutual attraction that we feel instinctively and now call love'. Importantly, 'the authors, alongside providing cogent evidence for 'emotional' and sexual instincts etc, also provide evidence that genes are not fixed instructions, but rather 'take their cue' from nurture/the environment' (Owen, ibid). Hamer and Copeland (ibid: 179) investigated whether there was a 'correlation between the D4DR gene and number of sexual partners' in male subjects. The molecular geneticists had previously established that 'novelty seeking' and sexual behaviour are linked, and that 'novelty seeking' is in part mediated by the D4 dopamine receptor gene. They found too that there was indeed a link between D4DR genes and the number of sexual partners in men, 'but in an unexpected way' (ibid). Looking first at heterosexual men, the geneticists found that the men with 'the long form of the D4DR gene, the high novelty seekers, had slightly more female partners than those with the short form, the low novelty seekers' (ibid). It should be emphasised that the trait referred to here as 'novelty seeking' means 'finding pleasure in new, varied, and intense experiences' (Hamer and Copeland ibid: 178). It was the study by Bogaert and Fisher (1995) at the University of Western Ontario which probably did more than any other towards establishing the idea that a novelty-seeking score was a better predictor of the number of sexual partners than other variables such as masculine age, physical attractiveness etc. 'The more a person was a thrill seeker, the more partners he had' (Hamer and Copeland, ibid). Hamer and Copeland's study became rather interesting when they asked how many other men had the respondents slept with. Despite the heterosexual orientation of the respondents, 'some had slept with another man, usually just once and when they were young' (ibid: 179). Here was a strong correlation to the D4DR gene. 'Straight' men with the long form of the gene, the high novelty seekers, 'were six times more likely to have slept with another man than those with a short gene', and 'about half of the long gene subjects had ever had a male sexual partner compared with only 8 percent of the short gene men' (ibid). The reverse was true for 'gay' men. As Hamer and Copeland expected, the homosexual respondents 'had more male partners than the straight men did female partners' and 'the D4DR gene had the expected effect' (ibid: 180). However, the effect of the gene 'was much stronger for the number of female partners of the gay men' (ibid). Those with the long, high novelty seeking form of the D4DR gene 'had sex with more than five times as many women as did those with the short, low novelty-seeking form'. Although, as the geneticists acknowledge, the gay men *may* have had sexual relations with women in part because of social pressure, 'it seemed that a desire for new experiences also played a role' (ibid).

According to Hamer and Copeland ….

> Those results show that the D4 dopamine receptor gene does influence male sexual behaviour, but indirectly. For a straight man, sleeping with another man is about as novel as you get. For a gay man, having sex with a woman is equally unique. Does this mean that D4DR is a "promiscuity gene" and that an errant husband can tell his wife, "I couldn't help it, it was genetic"? Of course not. A gene doesn't make a person commit adultery. It simply determines the way certain cells respond to dopamine, which in turn influences a person's reaction to novel stimuli. How a person reacts to that stimuli is more a matter of character than of temperament (ibid).

Arguably, Hamer and Copeland (ibid) provide further evidence that it is a mistake to engage in *genetic fatalism*; predisposition need not imply inevitability. Additionally, it is a mistake to deny genetic influences altogether in the form of instincts in the *oversocialised* manner of Gagnon and Simon (1973) and Giddens (1993). As previously stated in this Chapter, we are concerned here, at this point, with establishing whether we are justified in modifying Sibeon's original metatheoretical framework so that it entails acknowledging *biological variables* in any discussion of causality in relation to human behaviour. With this task in mind, we return to the approach of Ridley (1999, 2004), whom it will be recalled, favours an approach to the so-called 'nature versus nurture' conundrum, which can be called '*nature via nurture'*, in which genes and environment are accorded *plasticity* and *mutuality*.

In *Genome: The Autobiography of a Species in 23 Chapters*, Ridley (1999) discusses the relationship between genes, environment and freedom in some depth. He maintains that, 'the crude distinction between genes as implacable programmers of a Calvinist predestination and the environment as the home of liberal free will is a fallacy' (ibid: 303). He makes the compelling case that nature can be 'much more malleable than nurture' (ibid). The author claims that virtually all the evidence purporting to demonstrate how parental influences shape our character is flawed. He acknowledges the correlation between 'abusing children and having been abused as a child', but argues that the link can be accounted for 'by inherited personality traits'. Ridley goes on to argue that, 'the children of abusers inherit their persecutor's characteristics, and correctly controlled for this effect, social studies 'leave no room for nurture determinism at all' (ibid: 304). As Harris (1998) has shown, it is rare, for example, that the step-children of abusers become abusers. Taking this a stage further, Ridley (ibid) suggests that this tendency to ignore any possibility of *nurture determinism* characterises 'virtually every standard social nostrum'. Here he refers to the idea that 'criminals rear criminals' and so on.

Interestingly, Harris (ibid), after a career of writing psychology textbooks, 'suddenly began questioning them a few years ago' to find (according to Ridley, ibid: 5) that because 'virtually no studies had controlled for hereditability, there was no proof of causation at all in any study'. Correlation was being presented as causation, in other words. Yet as Ridley (ibid) suggests, 'from behaviour genetics studies, there was new, strong evidence against what Rich Harries calls 'the nurture assumption'. As Harris (ibid) has shown in her work, studies of the divorce rate of twins, as an example, reveal that genetics account for around half of the variation in divorce rate, non-shared environmental factors for another half and shared home environment for nothing at all. In other words, as Ridley (ibid: 305) suggests, 'you are no more likely to divorce if reared in a broken home than the average – unless your biological parents divorced'. Similarly, Harris's (1998) work reveals how studies of criminal records of adoptees in Denmark showed a powerful correlation with the criminal record of the biological parents and a very small correlation with the criminal record of the adopting parents (and that disappeared when controlled for peer-group effects), whereby the adopting parents were found to live in more, or less, criminal neighbourhoods according to whether they themselves were active criminals. Harris has, arguably, played a part in challenging what Ridley (ibid) calls, 'the dogma that has lain, unchallenged, beneath twentieth-century social science: the assumption that parents shape the personality and culture of their children'. As Harris (ibid) suggests, Freud's psychology, Watson's behaviourism and Mead's anthropology contain assumptions about *nurture determinism* that were never tested. Importantly, as Ridley (ibid: 307) has made clear, Harris's work 'lays bare just how much more alarming social determinism is than genetic'. Far from 'leaving room for free will, it rather diminishes it,' and 'a child who expresses her own (partly genetic) personality in defiance of her parents' or her siblings' pressures is at least obeying endogenous causality, not someone else's' (ibid).

As we have seen in the section of this Chapter entitled 'Building Bridges', it is a fallacy to equate genetic determinism and predisposition with fatalism. Yet, 'the myth persists that genetic determinism is a more implacable fate than social determinism' (Ridley, ibid: 308). As the author makes clear, it makes little sense to argue that 'biological determinism' threatens the case for political freedom. We have seen how Rose and Rose (2000: 107) accuse evolutionary psychologists of being part of a 'restored and re-energised neoliberalism' earlier in the Chapter. As Brittan (1998: 10) has argued, 'the opposite of freedom is coercion, not determinism'. Ridley (ibid) too argues that, 'we cherish political freedom because it allows us freedom of personal self-determinism, not the other way around'. For Ridley (ibid: 313), 'freedom lies in expressing your own determinism, not somebody else's', and it is 'not the determinism that makes a difference, but the ownership'. He goes on to argue that if freedom is what we cherish (and Rose and Rose (2000) appear to attack 'biological determinism' partly because they view it as constraining political freedom), then it is preferable to be 'determined' by forces that originate inside ourselves and not within others. According to Ridley (ibid), part of the public's revulsion at cloning; a biotechnological technique which we look at in the next Chapter, originates in the idea that what we see as uniquely our own could be shared by another being. Ridley goes on to argue that, 'the single-minded obsession of the genes to do the determining in their own body is our strongest bulwark against loss of freedom to external causes' (ibid: 313). The author makes clear that his optimism lies not in a 'gene for free will' (there is *no* single gene), but rather in the direction of ....

> Something infinitely more uplifting and magnificent: a whole human nature, flexibly preordained in our chromosomes and idiosyncratic to each of us. Everybody has a unique and different, endogenous nature. A self. (ibid).

Ridley makes a compelling case for the *plasticity* and *mutuality* of '*nature via nurture*', genes and environment. We have examined the evidence that suggests sexuality is, at least, partially genetic in terms of causality in the work of Hamer and Copeland (1999). Perhaps we should also examine, a little more closely, Ridley's (ibid: 104) claim that, 'The evidence that grammar is innate is overwhelming and diverse'. Hamer and Copeland (ibid) suggest that there is growing evidence that genes play a very important role in human language. One line of evidence comes from, 'the fact that language is so specific to our species' (ibid: 231). Just as human beings possess the greatest cognitive ability of all the species, so they have the greatest ability to communicate thoughts. There is, as the authors suggest, no animal parallel to our complex language that is able to distinguish past, present, and future and express concepts pitched at a very high plane of abstraction. It is not unreasonable then, for Hamer and Copeland (ibid: 232) to suggest that, 'there must be something about the human genetic blueprint that makes us especially capable of language'.

As the authors suggest, another line of evidence points to the very complicated rules found in all human languages. All languages have their own precise grammar, syntax and pronunciation etcetera. This is what prompted the linguist Noam Chomsky, according to Hamer and Copeland (ibid), to theorise that a newborn human baby's brain is 'prewired' with the basic ability to recognise sounds and to learn the rules of grammar and syntax. As the authors show, Chomsky postulated an "innate language acquisition device". According to Chomsky's theories, the specific language an infant learns depends completely upon the environment, but 'the basic ability to learn the rules and apply them to the billions of different combinations is genetically preprogrammed' (Hamer and Copeland, ibid).

Hamer and Copland suggest that one way to 'isolate' the specific genes involved may be to study what happens when the "language acquisition device" goes 'awry', as in the case of a family sharing a specific language impairment (SLI) which 'mangles' language. In such cases, the authors suggest, children start talking late, and then have great problems articulating words, and make grammatical errors which may persist through adulthood. Such people can have 'normal' intelligence and possess skills in many directions, except language. Specific language impairment, as Hamer and Copeland make clear, runs in families, which suggests the distinct possibility of a genetic basis. The work of Pinker (1994) describes a family in which the grandmother has the impairment, as do 4 of her 5 children. They in turn have 23 children, of whom 11 have the 'language problem'. However, as Pinker (ibid) makes clear, one of the grandmother's daughters speaks 'normally', and so do all their children, as if this 'side of the family' had escaped the gene. According to Hamer and Copeland (ibid: 233), 'This is just the sort of pattern expected for a condition caused by a single dominant gene, although the gene itself has not yet been identified'.

Regarding dyslexia and biological causality, more progress has been made in locating the genetic roots. Once called "word blindness", dyslexia is characterised by difficulty in learning to read 'despite adequate intelligence and character' (Hamer and Copeland ibid). As the authors suggest, it appears to be caused by a 'fundamental brain disconnect between different words and their meaning' (ibid). The malfunction is specific for reading, and as the authors show, dyslexics can be highly intelligent; they "just" cannot read adequately. As many as 8 per cent of children are 'reading disabled by standard school system criteria' (ibid). Children with dyslexia are also the largest group of students who receive 'special' education services, according to the authors. Dyslexia was first recognised as a discrete disorder in 1896, and 'within a decade it was known to run in families' (ibid). Since then, the authors make clear, family and twin studies have shown 'a substantial genetic effect' (ibid). Most dyslexics have, at least, one other person in the family who has reading problems, and the rate in identical twins is 40 percent.

Hamer and Copeland acknowledge that a reading disability is clearly a very complex disorder with many different causes, inherited and environmental. Even so, there appear to be some families in which one major gene is involved. As the authors make clear, scientists at the University of Colorado in Boulder (USA) and at the Boys Town National Research

Hospital started a 'gene hunt' in families with at least two dyslexics. They found evidence 'linking reading scores to chromosome 6 and found similar results in fraternal twins' (ibid). 'More recently, a different group of scientists at Yale University, found linkage to the same chromosome in yet another group of dyslexic families' (ibid: 233-234). As Hamer and Copeland suggest, these appear to be 'solid results', but it is not yet known which gene is involved or how it works. What does appear to be evident is that….

The gene has a major effect on the ability to break down long words into syllables, but not on comprehension of short words. That suggests something quite specific – a breakdown in the brain circuit used to segment words – rather than a more general problem. So when the actual gene is found it may tell us something fundamental about that particular part of intelligence dealing with language and reading (Hamer and Copeland, ibid: 234).

Although Hamer and Copeland claim that the 'gene for grammar' has yet to be isolated, more recent work by Enard et al (2002) and Lai et al (2001) suggests that a genetic mutation *may* be responsible for Severe Language Impairment (SLI). As Ridley (2003: 214) comments, 'It is the first candidate for a gene that may improve cultural learning through language'. We have previously seen how SLI has long been known to run in families from the work of Hamer and Copeland. According to Ridley (ibid), 'there is indeed a gene on chromosome 7, responsible for this disorder in one large pedigree and in another smaller one'. The gene is known as Forkhead box P2, or FOXP2 for short, and according to Lai et al (ibid), it is a gene whose job is to 'switch on' other genes – a transcription factor. When it is 'damaged', the person never develops full language.

Dunbar (1996) has suggested that language took over the role that grooming occupies in ape and monkey society – the maintenance and development of social bonds. As Ridley (ibid) shows, Dunbar (ibid) argues that when human beings began to live in large groups, it became necessary to 'invent' a form of social grooming that could be done to several people at once: language. Dunbar has noted that human beings do not use language simply to communicate 'useful' information; they utilise it in the main for social gossip. As Ridley (ibid: 218, 219) suggests, 'according to the scanty fossil evidence, speech, unlike manual dexterity, appeared late in human evolution'. As the author shows, the 1.6 million-year old Nariokotome skeleton discovered in 1984 in Kenya has in the region of the neck vertebrae space for only a narrow spinal cord similar to an ape's, and approximately half the width of a modern human spinal cord. As Walker and Shipman (1996) make clear, modern humans need a broad cord to supply the nerves to the chest for close control of breathing during speech. As Ridley (ibid: 219) shows, 'Other still later skeletons of *Homo erectus* have high ape-like larynxes that might be incompatible with elaborate speech'. Indeed, some anthropologists such as Tattersall (cited as 'email correspondence' by Ridley (ibid)) are tempted to infer that, since the attributes of speech appear so late in human evolution, language may have been a 'recent' invention, appearing as little as 70,000 years ago. Arguably, language is not the same thing as speech, and as Ridley (ibid) observes, 'Syntax, grammar, recursion and inflection may be ancient, but they may have been done with hands, not voice', and 'perhaps the FOXCP2 mutation of less than 200,000 years ago represents not the moment that language itself was invented, but the moment that language could be expressed through the mouth as well as through the hands' (Ridley, ibid).

It is at this point that we decide whether to modify Sibeon's original *anti-reductionist* framework to include focusing upon the *biological variable*, or evidence from evolutionary psychology and biological science for an, at least, in part genetic basis for human behaviour.

After considering the evidence that genes play a role alongside environment, in particular from Ridley (1999, 2003), Hamer and Copeland (1999), Enard et al (2002), Lai (2001), Dunbar (1996), Pinker (1994), Harris (1998), Bogaent and Fisher (1995), Curry (2003) and Cosmides and Tooby (1997), it is suggested here that there *is* sufficient evidence to warrant incorporating a focus upon the *biological variable* into the 'new' metatheoretical framework, alongside *agency-structure, time-space* and notions of Foucauldian *power*. We should keep Ridley's notion of '*nature via nurture'* firmly in mind when focusing upon *biological variables* in analysis, the 'feedback loop' which embraces genes *and* environment, acknowledging *plasticity* and mutuality. We need to remember, drawing upon Ridley (ibid) and the work of other evolutionary psychologists, that nurture depends on genes, and genes need nurture. Genes predetermine the broad structure of the brain of *Homo sapiens*, but they also absorb formative experience, react to social cues and need to be switched on, and external events – or free-willed behaviour – can switch on genes. We need to avoid *oversocialised* conceptions of the person (Gagnon and Simon, 1973; Giddens, 1993) and the tendency to associate genetic determinism/predisposition with inevitability or *genetic fatalism* (Newton's (2003) treatment of the work of Walter Cannon). We also need to focus upon *dualism* rather than the *duality* favoured by those of the 'embodied school' for example. This is to avoid what Archer (1995: 167-8) calls '*central conflation*'. Sometimes, in the course of analysis, it may become necessary to look at the 'biological' and 'the social' separately, whilst acknowledging links, for example, the point made by Ridley (ibid) that genes take their 'cue' from nature. Approaches which favour *duality*, such as those of Newton (ibid), Shilling (1993) and Freund (2001), make the separation of the two variables virtually impossible. This mirrors the *conflation* of *agency* and *structure* to be found in Giddensian *structuration* theory (Archer, ibid).

Having arrived at the decision to modify the framework, we may go on in Chapter Four to look at general issues of an ethical (particularly in respect of eugenic potential) and sociological implication pertaining to selected examples of human biotechnology and the controversial Human Genome Project. We look briefly at material from the 'sociology of science', considering, in the light of the Genome Project the recent transformations of science and its changing role within society (Bucchi, 2004). In addition to a focus upon issues pertaining to human biotechnology in general, we also consider the possibility that the '*Bio-medical model'* (Powell and Owen, 2005) may be *reductionist*. Whilst the case has been made in Chapter Three for the acknowledgement that 'genes do influence behaviour' (Ridley, ibid: 306), and for the avoidance of *genetic fatalism* and the *oversocialised* view which denies biological causality altogether, it is also the case that we should avoid reductionist 'Bio-medical' accounts which have developed, 'an all-encompassing range of bio-medical and totalised discourses', by which, 'many forms of social injustice could be justified as natural'' (Powell and Owen, ibid: 31). The emphasis here is upon what Ridley (1999, 2004) calls '*nature via nurture* applied to Sibeon's original *anti-reductionism*, not a case of either nature or nurture. As overemphasis upon genetic causality, which ignores social variables would be considered here as illegitimate as the *oversocialised* examples previously identified. Chapter Four, as previously stated, deals with an overview of general issues pertaining to human biotechnology, preparing the ground for an application of the framework developed in Chapters Two and Three to the study of selected examples of human biotechnology. The application of ideas from the framework is therefore not the main objective of Chapter Four. Nevertheless, some terms such as *reductionism*, *functional teleology, essentialism* and *reification* (Sibeon's original four 'cardinal sins') are applied to material, in particular that pertaining to 'bio-medical' models.

Chapter Four

Human Biotechnology: Selected Issues and Implications

In what follows, we examine selected examples of human biotechnology and the controversial Human Genome Project in terms of ethical (in the case of biotechnology, eugenic potential is highlighted in particular) and sociological implications. Some selected perspectives from the 'sociology of science' are considered, and with regard to the Genome Project in particular, how it 'exemplifies a number of ongoing patterns of change in contemporary science' (Bucchi, 2005: 134). Also considered is what Powell and Owen (2005) identify as '*the Bio-medical model'*. Whilst it is essential to avoid *oversocialised* perspectives which deny biological causality altogether (the work of Gagnon and Simon (1973), and learned sexuality' is a case in point), it is also essential to avoid *reductionist* 'Bio-medical', totalised discourses which deny environmental causality and emphasise 'rigid explanations of human behaviour' (Powell and Owen, ibid: 27). The point is emphasised here that, whilst it is essential to acknowledge genetic causality, it is also essential to acknowledge how genes 'take their cue' from nature. In the spirit of Ridley's (1999, 2003) work, the *'nature via nurture*' rather than nature versus nurture model is favoured here, and in the metatheoretical framework. Lynn (2001), discussing developments in genomics in the light of the Human Genome Project, points out that there are six techniques of biotechnology that have been used to promote eugenics, or that have the potential to promote eugenics. These are *artificial insemination by donor (AID); egg donation; pre-natal diagnosis of fetuses with genetic diseases and disorders and termination of the pregnancies where these are identified; embryo selection, consisting of the production of genetically identical copies of individuals*; and *genetic engineering,* used here in the sense of the insertion of new genes. The Chapter examines the six, selected examples after we have considered some significant implications of the controversial Genome Project in some detail.

The Human Genome Project

The Human Genome Project is an international research programme that was created in order to characterise the genomes of human beings and other organisms; to develop the technology required to do so; and to address the ethical, legal and social implications of the new information (Dennis and Gallagher 2001). The possibility of sequencing the human genome was first discussed in the mid-1980s (Sulston and Ferry, 2003). The HGP officially 'started' in 1990 with a 15 year plan to map and sequence the human genome. The project was launched with the aim of developing genomic tools and resources, and of mapping and sequencing the human and other genomes. Throughout the project, maps of human chromosomes have been created; sequencing technology has been greatly improved; computational tools have been designed; and the strategies for gathering, analysing and storing genomic data have been developed (Davies, 2001). The study of the effect on society of large amounts of 'new' genetic knowledge and technology, and the recommendation of policy to maximise the benefits and reduce the risks, have been an integral part of the HGP since its inception (Bucchi, 2004; Sloan et al 2000).

Initial efforts were focused upon the genomes of laboratory organisms such as yeast and the roundworm. As Dennis and Gallagher (ibid: 31) suggest, 'these projects helped to refine DNA – sequencing technology and the conceptual strategies needed to tackle more complex genomes, and at the same time provided essential information for interpreting the human sequence'. Only a very small amount of the human genome encodes proteins. Comparing human DNA with that of other organisms helps to identify human genes amongst vast numbers of sequences, and serves to provide clues as to their functions.

Dennis and Gallagher (ibid) discuss how, in 1996, a series of pilot sequencing projects was initiated to test the feasibility of deciphering the human genome. Within a couple of years, these efforts were viewed as successful, and the large-scale effort to sequence the human genome was launched in March 1999. Around this time, the HGP adopted a strategy that would first generate a 'working draft', in order that genomic scientists would have useful data covering the entire genome as quickly as possible, rather than having to wait for the sequence to be 'finished'. With the infrastructure and resources in place, productivity 'snowballed', and more than 90 per cent of the sequence was produced in just 15 months. In February 2001, HGP scientists published the sequence, assembly and analysis of the 'working draft' of the human genome. As Brown (2000) suggests, the HGP's approach has been described as painstaking and precise. Beginning with blood and sperm cells, the 'public' team separated out the 23 pairs of chromosomes that hold human genes. Scientists then clipped bits of DNA from every chromosome, identified the sequence of DNA bases in each bit, and, finally, matched each snippet up to the DNA on either side of it in the chromosome.

In contrast, their 'private' rivals, Celera, took a shorter route, 'shredding the encyclopaedia all at once' (Brown ibid: 41). Celera's so-called 'shotgun sequencing' tears all the genes into fragments into a whole genome. As Brown (ibid) suggests, the HGP and Celera teams disagreed over what makes 'a finished genome'. In Spring 2000, Celera announced that it had finished sequencing the rough draft genome of one anonymous person (rumoured to be Craig Venter, the 'head' of Celera) and that it would sort the data into a map in just six weeks. However, as Dennis and Gallagher show, the 'public' HGP team immediately objected, with Francis Collins noting that Celera fell far short of its original genome sequencing goals. In 1998, when the company began, Celera scientists planned to sequence the full genomes of several people, checking its 'consensus' genome 10 times over. In its April announcement, however, Celera declared that its rough genome sequencing was complete with just one person's genome, sequenced only three times.

Although many news accounts have characterised the HGP and Celara as competing in a 'race', (Dennis and Gallagher, ibid), the company, according to Brown (ibid), had a decided advantage. Because the 'public' HGP *is* a 'public' project, the team of scientists routinely 'dumped' all its data into *Genbank,* a public database available through the Internet. Like many other companies and individuals, Celera has used that data – in its case, to help check and 'fill the gaps' in the company's 'rough-draft' genome. In essence, Celera utilised the 'public' HGP's genome data to stay one step ahead in the sequencing 'race'. As Dennis and Gallagher (ibid: 33) have noted, 'in a joint announcement on 26 June 2000, the HGP and Celera celebrated their respective first assemblies of the human genome'. Celera went on to publish its initial analysis in the 16 February 2001 issue of *Science*, the same week that the 'public' HGP reported their findings in *Nature*.

Opposition to the Genome Project from the perspective of rational resource allocation was expressed by many critics at the outset of the HGP (Rechsteiner 1991; Lewontin 1992; Tawber and Sarkar 1992). In a historical perspective, Sloan et al (2000) argue that the history of the Eugenics movement in the early decades of the 20th century presents an ever-present concern about the practical social effects of a 'new' form of eugenics (Kevles 1985; Paul 1998; Pernick 2000; Caplan 2000; Bucchi, 2004 and Wilkie 1994). Sloan et al (ibid) argue that a tendency to accentuate only 'positive' benefits of the HGP in the official justifications of the 'public' project greatly concerns observers such as Dreger (2000). Humanistic scholars generally have reasons to pay attention to the HGP. Applied ethicists 'must take account of the issues now confronting professional genetic counsellors, lawmakers, insurance companies and medical legalists that directly result from the production of such broad genetic information about human beings' (Sloan et al ibid: 4). The information generated by the HGP has consequences for medical insurability, the equality of access to health care, and the practices of genetic counsellors (Brown and Marshall 1993; Association of British Insurers, 1997; Pokorski, 1997; Timson, 1997).

Another set of problems, according to Sloan et al (ibid), is more theoretical, involving critics with theoretical interests that might not ordinarily take part in current discussions about health science and bioethics professionals. Scholars concerned with 'science studies' in a broad sense (history, philosophy, sociology etc) will find in the HGP a field of scientific development that displays all of the issues involved in an understanding of contemporary science and technology. As Lenoir and Hays (2000) and Beatty (2000) have respectively acknowledged, the HGP is 'Big Science' in the fullest extent of the term, involving what Sloan et al (ibid: 5) calls, 'the complex interplay of scientific theory, technological manipulation, governmental politics and industrial capitalism'. Arguably, as Wade (1998), Beatty (ibid) and Dreger (ibid) have noted, although the project was initially conceived with very broad international ideals, it has been shown that it can be quickly 'balkanised' by competing commercial and national interests. Opitz (2000) makes the point that human genetics is a science that carries immediate social and ethical consequences, accompanied by very strong ideological components which are easily masked by language and hidden background assumptions. He refers specifically to the 'cascade' of genetic information received by the public every week through the media, arguing that this information carries with it a pervasive sense of reductionism of the human to the biological and a belief in genetic determinism. Opitz appears to engage in *genetic fatalism* here, as the phrase 'genetic determinism' is used in a 'negative' sense, implying inevitability. Where such 'genetic reductionism' is leading contemporary Western culture concerns authors such as Pernick (ibid). His discussion of the history of the 'popular' Eugenics movement and the 'popular' understanding of genetic defects supplies reasons for reflection on the issue as it effectively illustrates how a 'popular' understanding of science can influence public discourse, and even filter down into school curricula.

Various authors have addressed the question of Eugenics surrounding the HGP and genetic research in general. Caplan (ibid) confronts directly the fears that a 'new' eugenics might have the same kind of unforeseen consequences that accompanied the original 'state' eugenics movement, and seeks to allay such fears. Kitcher (2000) explores the concept of 'Utopian' eugenics within a framework that develops upon the concept of rational advantage and social optimality considerations.

Wilkie (1994) made what was possibly one of the first significant contributions towards an explanation of how the Genome Project might serve human needs, and the possible ethical problems to which the discipline of molecular biology genetics gives rise. (Wilkie (ibid: 169) refers to the 'changes' in 'scientific practice' that will be brought about by the Human Genome Project as amounting to, 'one of Kuhn's lesser revolutions when compared to developments such as Darwin's theory of evolution', though he makes the point that, 'in the realm of social affairs the Human Genome Project has the potential to transform our society and our view of our own place in the scheme of things'. The reference to Kuhn pertains to the work of Thomas Kuhn (1962), the American philosopher of science who characterised his discipline as, 'an activity in which periods of 'normal' science are interrupted by the occasional revolution' (Wilkie, ibid: 167). Wilkie refers to the many practical consequences for, 'seeking medical care, would-be parents planning a family, prudent investors seeking to save for a pension in their old age' through to 'insurance companies trying to access the actuarial risk of giving someone life assurance, and for prospective employers assessing the health and capabilities of their workers' (ibid: 170).

Some changes and consequences, according to Wilkie (ibid) will be subtle and unexpected. He provides the example of how, as individual human genes are located and analysed 'most of them will be found to be similar to genes already known from other animals' (ibid: 171). It will then, according to the author, become, 'inescapably clear just how closely related human beings are to the rest of the animal kingdom' (ibid). This may generate unease amongst the 'lay public', despite the fact that our common heritage with the rest of the animal kingdom has been commonplace with biologists since Darwin's day. 'Once the Human Genome Project is complete' the close relationship between humans and the rest of the animal kingdom will be demonstrated', 'not just at the level of comparative anatomy of the whole animal but through the comparative anatomy of their and our genes as well' (ibid). Wilkie poses the question of whether such genetic developments will cause us to rethink and reassess how we use animals for our own ends and perhaps to conclude that many things which we currently do to animals are no longer morally acceptable? (ibid). The question of whether reminders of our animal ancestry will serve to reduce the respect human beings have for each other's humanity is also posed (ibid). Wilkie goes on to identify a second consequence of the HGP, which (paradoxically) may be a tendency to stress the differences between one human being and his or her neighbour (ibid). Wilkie makes the point that it is commonplace that differences of hair and eye colour are genetically determined, and that few put much stress upon the colour of an individual's eyes. However, there are other genetic differences, such as skin colour, which have become associated with enormous social and economic differences (ibid). Wilkie stresses that other human traits such as intelligence and sexual orientation have been claimed as genetically determined (although the evidence is controversial in the extreme) and those who do not conform to the social norm or expectation are severely disadvantaged and discriminated against in many societies (ibid).

Another possible consequence of the HGP, 'stems not from any likely discovery, but from the project's very existence' (ibid: 172). Wilkie here warns against the possibility that we may take an increasingly atomistic view of human beings and indeed of life itself (ibid). Under the great impact of ever more new information and discoveries about human genetics, we may come to define ourselves and our lives in genetic terms and neglect the rest (ibid). Wilkie warns against the danger that, 'we may all become reductionists', and although he does not mention the environmental influences upon human behaviour, they are implied by the use of his phrase, 'the complexity and richness of life' in reference to variables 'missed' when 'holistic' approaches are abandoned (ibid). As previously emphasised, the metatheoretical framework developed in Chapters Two and Three, applied on occasion in this Chapter and in greater detail in Chapter Five, also emphasises an avoidance of *reductionism*, suggesting instead that *both* genetic and 'environmental' causality play a part in explanations of human behaviour.

Bucchi (2004) looks closely at the implications of the Genome Project, and in particular, how it represents various patterns of change within contemporary science, which will be explained and examined in what follows. Bucchi's (ibid) work covers how the social and physical world has been shaped by science, and the relationship of the human species to it. Bucchi summarises the major theoretical positions within what has become known as 'the *sociology of science'*. These include several 'schools' of thought and sociological approaches. Kuhn's (1962) work upon science and revolutions is examined. Kuhn (ibid) sets out a theory of scientific change grounded in concepts such as 'paradigm' 'revolution' and 'normal science'. According to Kuhn, 'science does not advance smoothly along a linear path and by gradual approximations to the truth; rather, it is characterised by abrupt 'leaps' and profound 'discontinuities' – revolutions in a word' (Bucchi, ibid: 26). Such 'discontinuities' interrupt periods of 'quietness' characterised by the conduct of 'normal science'. Kuhn (1962: 10) defines 'normal science' as ....

Research firmly based upon one or more past scientific achievements, achievements that some particular scientific community acknowledge for a time as supplying the foundation for its further practice.

This outcome, as Bucchi (ibid) explains, orients the work of scientists operating in a particular field or sector, and particular period in time, on the basis of consensus among the 'community'. Disputes may occur, but there is no disagreement on fundamental issues. Kuhn has given the name 'paradigm' to this result/group of results. As Bucchi (ibid: 27) points out, 'an example of a recently and solidly founded paradigm is the Big Bang theory, or the idea that the universe originated in a specific event (a 'singularity'). Bucchi (ibid) looks also at 'the birth of the sociology of science', in particular the work of Merton (1952, 1968), Eltzinga and Jamison (1995), Mannheim (1925) and Miller (1994). Reference is made to the studies by Merton and colleagues of 'the organisational and functional aspects of science as an institution capable of self regulation' (Bucchi, ibid: 16). The approach is most famous for the description of the normative structure of science, in other words which norms and values, according to Merton (1942) guarantee the functioning of science. Merton (ibid) identified *universalism* (scientists are rewarded exclusively on the basis of the results obtained, regardless of social class, race or religion); *communism*: results/discoveries belong to the scientific community and society at large; *disinterestedness*: researchers pursue the primary objective of knowledge progress, achieving recognition indirectly; *organised skepticism*: all researchers have to be willing to evaluate all hypotheses critically, including their own until confirmations are obtained. Merton's 'institutional' approach, 'diluted into subsequent 'schools' of sociology of science' with general sociological theories and concepts (Bucchi, ibid: 22).

The author looks too at 'the Edinburgh school' otherwise known as the '*sociology of scientific knowledge*; drawing upon those such as Shapin (1982), Bloor (1976), MacKenzie (1978) and Keller (1995). Proponents of the SSK have examined the relationship between society and science from various perspectives but, according to Bucchi(ibid: 48-49) they are often identified with 'the strong programme' consisting of 'methodological principles for the sociological analysis of scientific knowledge' which epitomise 'the sociology of science', such as *causal, impartial* and *reflective* analysis. Bucchi's (ibid: 61) analysis extends to attempts to 'resolve the difficulties of the strong programme' which, rather than setting scientific theories in relation to specific social/historical contexts, 'delve into the process itself that leads to the theory's formulation; (ibid: 61-62), placing them under great focus. Bucchi (ibid: 62] identifies a 'merging' of a 'certain distancing from naturalism and positivism' which is 'apparent in the Edinburgh school, and in the strong programme' with ethnomethodology. Here, scientific knowledge is not only socially conditioned but, from the very beginning, 'constructed and contributed through microsocial phenomena' (Latour and Woolgar, 1979: 236). Bucchi (ibid: 70) goes on to look at actor-network theory, as 'an attempt to expand the explanatory capacity of the microsociological approaches to science'. For the proponents of such theories, such as Latour (1987), and Callon and Latour (1981), 'science has two faces, like the Janus of Roman mythology: on the one hand there is 'ready-made' science; on the other, science 'in the making' or 'research' and whilst it is the task of epistemology to analyse the qualities of the former, 'it is the task of the sociology of science to study the latter'. It will be recalled that actor-network theory was referred to earlier in Chapter Two, with reference to the persuasive arguments of Callon and Latour (ibid) to the ends that power has no single source, or prime cause. However, it will be also recalled that, *'contra'* those such as Callon and Latour and Foucault, 'power can be 'stored' in positions/roles and in social systems' (Owen, 2005: 23). Aspects of Foucauldian power are incorporated within Sibeon's original framework, and in the modified framework employed here, which 'recognise the *dialectical* relationship between *agentic* and *systemic* forms of power' (Owen, 2006a: 193). In Chapter Five, actor-network theory is looked at in greater depth, alongside the related *posthuman agency* concepts and the framework is 'applied' for this purpose. Bucchi's (ibid) work moves on to cover the field known as STS – Science and Technology Studies, looking at the work of Barnes and Shapin (1979), Layton (1988), MacKenzie and Wajcman (1999), Bijker (1995) and others, and tracing the 'schools' development from Merton's (1938) study on 'Science, Technology and Society in 17th century England' to analysis of technology and trust (MacKenzie, 1996) in relation to warfare, and in particular material pertaining to the 'Gulf War' (Collins and Pinch, 1993). Before examining the implications of the HGP, Bucchi (ibid) finally considers the work of Sokal (1996) in a focus upon 'science wars'. Sokal famously penned 'Transgressing the Frontiers: Towards a Hermeneutic Interpretation of Quantum Gravity', which was published by the journal, *Social Text*. As Bucchi (ibid: 93) makes clear, 'the article was, in fact, an entertaining parody of a certain academic style of writing', was a mishmash of gibberish on physics and mathematics' and included a 'hoax experiment'. Sokal has declared that he wished to combat a currently fashionable postmodernist/poststructuralist/social-constructivist discourse – and more generally a penchant for subjectivism. The work of Sokal and Bricmont (1997) targets *relativism* as does the original framework of Sibeon and the modified framework employed here, although the authors do not suggest adopting ontologically-flexible, *meta-theory*, which is favoured here. As Bucchi (ibid: 126) suggests, the Human Genome Project is ....

> A project that lies at the core of significant transformation in the ways in which scientific knowledge is produced, distributed and utlised, and especially in the role played by biology in scientific research.

Earlier, in Chapter Two, we examined the Project in relation to Foucauldian *power*, and the argument was made for the Project as an example of Sibeon's (1996: 52 'emergent power' (Owen, 2006b: 911). It may be recalled that Owen (ibid) makes clear that the HGP is *not* an actor, but rather '*a supra-national organisation*', with the International Human Genome Sequencing Consortium (HGSC) as a '*policy network*' (ibid). It is also as Bucchi (ibid) puts it, 'one of the most ambitious enterprises', with the 'widest social implications ever undertaken in the history of science'.

Bucchi (ibid: 127) shows how, after the military victory over Japan, 'the American authorities immediately decided to develop a programme of scientific collaboration with the country that had only just previously been their enemy'. A priority of the programme was the study of mutagenic effects of nuclear radiation – 'presumably in an attempt to heal the wounds caused by the bombing of Nagasaki and Hiroshima' (ibid). Soon, genetics proved essential for the understanding of these radiation mutations and, as a consequence, 'the US federal agency responsible for nuclear programmes, the Department of Energy (DOE), found itself playing a leading role in biological research' (ibid), rather unusually. As Bucchi makes clear, on the other hand, as tensions between the two blocs appeared to be subsiding after many years of the so-called Cold War, there, arose a perception of Japan as a threat to American economic supremacy, and it was realised that competitiveness could be restored to the economy by focusing upon scientific and technological development, one of the great strengths of its rival, Japan.

Robert Sinsheimer, a molecular biologist on course to become Chancellor of the University of California at Santa Cruz, found that 36 million dollars had been allocated to the building of an optical telescope at his university. As Bucchi shows, at the same time, Santa Cruz was being considered as a possible site for a huge particle accelerator, the Superconductor Supercollider, which would have cost several millions of dollars, and Sinsheimer was prompted to ask why this money should not be given to biology. So Sinsheimer organised a meeting with leading molecular biologists in the US, for the purpose of examining the possibility of applying for funds to establish an institute for the study of the human genome at Santa Cruz. According to Bucchi (ibid), some scientists, 'immediately voiced their concerns over the exorbitant costs of the project' especially regarding the 'still inadequate' state of knowledge on the human genome. Estimates made at the time of the meeting apparently envisaged 15 years of work and 3 billion dollars of expenditure, with the fear that the project would remove funds from the research projects. Bucchi's view is that here we have a case of biologists 'accustomed to the style of Big Science', and 'afraid of losing their independence' as, unlike physicists, they did not have a, 'well-established tradition of international collaboration and the division of labour into small sub-projects' which could bring in students/postgraduates (ibid: 128). Further concerns were expressed about the social and political implications of the project in the sense that, 'information about a person's genetic make-up might lead to discrimination' by employers or insurance companies, or lead to 'eugenic policies historically associated with the memory of Nazism' (ibid).

As Bucchi shows, the Department of Energy quickly profited from the hesitations of scientists. The then director of the DOE, Charles Del Lisi, 'saw genetic research as justifying the existence of institutions like his well into the future' (ibid). De Lisi pledged the DOE's commitment to the human DNA sequencing project at a conference, held at Santa Cruz. The DOE possessed well-equipped research facilities and was able to 'direct funds out of its large reserve for nuclear research – five and a half million dollars in 1987 – to the project' (ibid). De Lisi maintained that the power of the DOE's computers (originally developed/tested for physics) in terms of calculation would be essential for genomic research, and reminded his peers that the two main nuclear research centres (Los Alamos and Livermore) had been at work upon a genetic library for a number of years.

Bucchi (ibid) describes how it did not take long for the scientific community to voice its hostility. In particular, the relatively 'high profile' Nobel prize-winner David Baltimore and James Watson (joint winner of the 1962 Nobel prize with Francis Crick, for the discovery of DNA structure), were convinced of the importance of the HGP, but, objected to its being controlled by the DOE bureaucracy. As Wilkie (1993: 81) suggested, perhaps genome research should be guided 'by the perceived needs of science?' As Bucchi (ibid) explains, the solution was to develop the project under the auspices of the National Institutes of Health (NIH). Watson intervened (now a scientific and public celebrity) and in 1988 was appointed director of genome research at the NIH with a budget of almost 30 million dollars (ibid). The year after, found Watson in charge of a National Center for Genome Research with an allocation of 60 million dollars. Bucchi points out that the development of similar projects in Britain, France etc, 'persuaded genome researchers that their efforts must be coordinated on an international scale' (ibid: 129). In 1988, an association known as HUGO (Human Genome Organisation) was created for the purpose of allotting tasks amongst the many laboratories. In 1990, the Human Genome Project itself followed, which can be regarded as an 'international consortium' (ibid), comprising 16 laboratories, a budget of around 3 billion dollars (two billion of which were provided by the US government). However, as Bucchi shows, this did not serve to dispel doubts amongst some scientists, and in 1991, a group headed by the members of the Department of Molecular Genetics at the Harvard Medical School wrote a letter to the journal *Science* in which they argued that there was no justifiable reason for devoting so much money to the genome project. Kevles and Hood (1992) mention how the group stressed that 90 per cent of the human genome was of no interest, and accused the senator sponsoring the HGP (Pete Domenici of New Mexico) of only doing so to favour his Alamos laboratories situated in his state. We have already covered the rivalries between the 'public' project and Celera Genomics, which does not require re-emphasising.

Bucchi refers to the discovery in 1980 by Michael Gottlieb (an immunologist working at the University of California at Los Angeles Hospital) of, 'a distinctly odd syndrome', in which (there were five cases under Gottlieb's observation) homosexual patients, 'besides displaying symptoms like diarrhoea and weight loss' had also developed, 'a rare form of pneumonia' (ibid: 131). Two of the patients died shortly after. Sensing that he was in the presence of an emergency situation, Gottlieb contacted an editor at the *New England Journal of Medicine* (a prestigious US medical journal) explaining the gravity of the situation, and asking 'how long the journal would take to publish a report warning of the emergency' (ibid). The editor, according to Bucchi, informed Gottlieb that it would require at least 3 months, during which time Gottlieb must refrain from making statements to the media (if he did, publication would be cancelled), and in any case publication could not be guaranteed. He advised Gottlieb to contact the Centers for Disease Control. As Bucchi shows, Gottlieb prepared a paper entitled 'Pneumocystis Pneumonia in Homosexual Subjects – Los Angeles', and sent it to the CDC for publication in their weekly bulletin on disease and mortality, the *Morbidity and Mortality Week Report*. There were many discussions at the CDC on the 'advisability' (ibid) of publishing the account, in particular because of the references to the homosexuality of the subjects affected. Meanwhile, several more cases of the syndrome were reported. A correspondent of the weekly report claimed that, 'numerous doctors in New York had come across similar cases but probably did not want to talk about them because they had articles forthcoming in medical journals' (ibid: 131-132). The gay daily *New York Native* gave prominence to the cases, but 'the local health district had hastened to deny the story' (ibid: 132). Finally, Gottlieb's report was published on an inside page of the *Morbidity and Mortality Week Report*, but with all references to the homosexuality of subjects removed.

As Bucchi (ibid) suggests....

> This minor episode, which marked the beginning of another scientific affair of great public concern, namely AIDS, highlights the three cardinal points around which the communication of science rotates: secrecy (to protect one's discoveries and prevent their plagiarism by other researchers), discussion of ideas among colleagues and the sharing of these ideas with the general public and their adaptation to the social and political context. A perhaps idealised conception of the scientific enterprise and its relationship with other social dynamics has long conceived these cardinal points as arranged in a sequence: the solitary and confidential development of theories and experiments, discussion with other specialists, and then dissemination of the results through the media and educational institutions.

Bucchi goes on to explain how the development of scientific activity and the huge changes that have taken place in its organisation and the role of variables usually considered external to it, such as pressure groups and the mass media, mean that there is intersection, tension and possibly conflict between the three points of the triangle. Arguably, the Genome Project demonstrates that 'simultaneously at work' may be 'pressures to keep a discovery secret, to obtain the cooperation of competing scientists, and to publicise the discovery in order to inform the public or simply to gain visibility, legitimacy and, in the end, economic resources' (ibid). Additionally, there is the impact of the HGP on the public imagination. Bucchi refers to the various metaphors used to describe the genomic undertaking: 'The book of life', 'map', 'the instruction manual for our species' etc. As the author states, the Project has incorporated a great deal of public resonance from its beginning. Considerations of political and social expediency hampered its acceptance by the scientific community. Thereafter, 'social scientists and philosophers were included in the Project' to examine ethical and societal implications (ibid). Conversely, 'those directly involved in the Genome Project have widely exploited its viability' and the 'association, emphasised mainly in the public domain, between genetics and the treatment of diseases' has ensured the HGP's overall success and wide acceptance despite opposition from 'numerous specialists' (ibid).

In the course of the Project's history there has been a conflict between transparency and secrecy. As Bucchi (ibid: 133) shows, the then President Clinton and the then Prime Minister Blair recommended that 'the entire map should be released into the public domain'. As the author suggests, the public consortium consequently accelerated its work schedule to move this release forward and compel Celera to do the same. Whilst there are obvious problems and dilemmas pertaining to the issues of the private character of genetic knowledge and secrecy the unfiltered dissemination of scientific results could also prove problematic. There is the issue of possible discrimination by employers and insurance companies (previously mentioned), and problems concerning the handling of such information which have already arisen in the therapeutic field. Bucchi cites the example of British media coverage of the discovery of a gene believed able to cause breast cancer – 'even if in only 5 to 10 per cent of cases' (ibid) which led to numerous women with cases of the tumour in their families undergoing prophylactic mastectomy. This conflict between transparency and secrecy has arisen at other levels too. In 1988, James Watson 'caused consternation' (ibid) by proposing that part of the work on the human genome should be allocated to the Soviets. Bucchi (ibid) shows how, in 1991, for two months during the Gulf War, the US Department of Trade 'blocked access to the NIH computers by foreign researchers collaborating on the Project for fear that information might be used for biological warfare'. Wilkie (1994) has made the same point.

According to Bucchi (ibid: 134), 'the story of the Genome Project exemplifies a number of ongoing patterns of change in contemporary science'. For the author, the Project was created 'for political purposes', which is 'typical' of post-war research policy, and developed 'in accordance with profound changes in the organisation of scientific research; a 'post-academic' science' (ibid). This 'post academic' science is characterised by the intersection of disciplines, by the heterogeneity of the actors involved, and by reflexivity and ideas pertaining to social accountability (Gibbons et al, 1994; Ziman, 2001). In organisational terms, the HGP 'can be viewed as in some way equivalent to the Manhattan Project for physics', representing a 'quantum leap' from a 'discipline of middling importance' to one which epitomises the 'new' 'Big Science' on a collective endeavour 'carried forward by huge and complex organisations' (Bucchi, ibid: 134). The Project represents a new transformed kind of 'Big Science' which has ousted the old 'military-industrial complex', replacing it with a new 'academic-industrial-governmental' complex (ibid, see also Etzkowitz and Webster, 1995) in which venture capitalists set the research agenda, researchers become akin to entrepreneurs and politicians mediate between them.

Nowotny and Taschwer (1996) suggest that the new 'Big Science' symbolised by the Genome Project differs from its predecessor in the sense that it has to constantly negotiate its importance and social acceptability with a wide variety of different contexts and publics. In support of their position, Bucchi (ibid: 139) comments that 'paradoxically, in fact, the increasing economic importance of science has induced society to demand a legitimation of science which goes beyond its capacity to produce development and material affluence'. Despite the *reification* (illegitimate attribution of *agency* to entities which are not actors) in both the positions of Nowotny and Taschwer (ibid) towards 'Big Science' and Bucchi (ibid) with regard to society, there is some support for such views also in the recent work of Koski (2005). Koski (ibid) in the December 2005 issue of *New Genetics and Society*, goes on to show how an examination of the development of the HGP in the United States helps to establish why and how the general public has become a 'major player' in science policy during the past quarter of a century, in particular, with regard to the ethical, legal, and social implications of research on human genetics. The attribution to science of greater 'social accountability' and the utilisation of 'scientific expertise to control research and development activities' and 'the creation of citizen panels where non-experts sit side by side with scientists and policy-makers' are some of the very 'visible' results of the process of increasing public attention/participation (Bucchi, ibid). In fact, as the previous author suggests, the increase in complexity and uncertainty or 'the emergence of a new economic rationality' (ibid) have invested society in general and science, with the consequence of what Nowotny et al, (2001) refers to as the increasing difficulty in establishing demarcation between the two. Haraway (1997) suggests that the new situation represents a 'New World Order Inc' in which biotechnology is the 'promised land', and knowledge, private property, industry and scientific research intersect, rather than there being merely a 'transformed science'. For Haraway (ibid), the OncoMouse ™, the first *patented* animal in the world, represents technoscience as a whole and what Bucchi (ibid) calls, 'a new, colossal, scientific revolution in which transgenic animals perform the same role as transurance elements during the Cold War'.

Risk And Governance

As this point it may be useful to reiterate that the intention in this book is to develop and apply a metatheoretical framework which can be used to contribute to the study of human biotechnology and the Human Genome Project. It is *not* the intention here to explore issues pertaining to biotechnology in general, or to do with *risk* and *governance*. In what follows, the main focus is firmly upon the six techniques of human biotechnology which are linked to reproductive processes and have been identified as having the potential to promote eugenics (Lynn, 2001). As Lynn (ibid) has shown, the six techniques identified at the beginning of the Chapter (including gene therapy, egg donation etc) are far more closely associated, in terms of possible implications, with eugenic issues than with any other issues. Nevertheless, it is necessary, having covered the Human Genome Project, to very briefly outline the literature pertaining to *risk* and to *governance* in relation to biotechnology, before proceeding to examine the eugenic potential of the six techniques of human biotechnology.

Regarding the governance of genomics in the late 20th century and early 21st century, Gottweis (2005) sketches the main features and modes of operation and tactics of an emerging genomics apparatus. The author shows how genomic governance is characterised by the process of the stabilisation of 'knowledge regimes'. Reference is made to a heterogenisation and globalisation of the actors and knowledge-generating systems in governance. Also emphasised is the transition of governing via risk to governance by uncertainty. Jones and Salter (2004) show how the collection of practices commonly known as 'biotechnology' poses a challenge to traditional mechanisms of regulation in the fields of science/technology. They emphasise that 'public trust' is a key measure of political success or failure. They go on to explore policy discourse analysis as a vehicle for exploring the politics of the relationship between human genetics governance and public trust. Martin and Frost (2003) suggest that genetic testing promises to revolutionise health care, and the authors go on to examine the way in which genetic testing has been governed and regulated by the medical profession and the state, analysing unresolved issues for public policy.

As Hannigan (1995) shows, sociologists of *risk* generally suggest that while risk is a socio-cultural construct, it cannot be confined to perceptions and social constructions alone. Technical risk analysis is an integral component of the 'social processing 'of risk (Renn, 1992). Dietz et al (1997) observe that the main themes in the sociology of risk have followed three separate but complementary directions, bound together by an emphasis on the social context in which decisions about risk are made. First, sociologists are concerned with how perceptions of risk differ across populations facing different life chances, and whether the 'framing' of choice stems from power differences. Second, sociologists of risk propose a model that 'reconceptualises the problem of risk perception by taking into account the social context in which human perceptions are formed' (Hannigan, ibid: 95). Third, risks, especially technological risks, are often conceptualised as 'components of complex organisational systems' (ibid: 96). This is exemplified by the work of Perrow (1984), a well-known analysis of 'normal accidents' in which an estimated probability of failure is built into the design of technologies with high risk potential.

Recent literature pertaining to *risk* in relation to biotechnology, includes work by Irwin and Lynne (2003), Beck [1992, 1999, 2002]; Durant and Bauer et al (1998); Lupton (1999); Revelli (2003); Ericson and Doyle (2003); Russell and Vogler (2001); Brodwin (2001); Van Loon (2002); and Bromley and Segerson (1992).

HUMAN BIOTECHNOLOGICAL DEVELOPMENTS

Lynn (2001) suggests that there are six techniques of biotechnology that, in his view, have been used to promote eugenics, or that have the potential to promote eugenics in the future. These are artificial insemination by donor (AID); egg donation; pre-natal diagnosis of foetuses with genetic diseases/disorders and termination of the pregnancies where these are identified; embryo selection, which consists of the 'ascertainment of the genetic characteristics of the embryos grown in vitro and the selection for implantation of those with genetically desirable characteristics (ibid: 245); cloning, consisting of the production of genetical identical copies of individuals; and genetic engineering, in the sense of the insertion of new genes' (ibid). We address the actual and potential development in what follows. At this point it must be emphasised that Lynn's highly controversial views on 'race' and geographical variations in intelligence are neither endorsed nor discussed here as they are irrelevant to the purposes of the book.

Artificial Insemination

This, as Lynn (ibid: 246) points out, 'entails taking semen from a donor and using it to fertilise a female'. The technique is used regularly by animal breeders, who take semen from males with desirable characteristics, with the desire that these will be transmitted to the offspring to produce genetically improved strains of livestock. During the 20th century, AID became a standard treatment for couples who wanted children but the husband was infertile, and by the end of the century thousands of children had been conceived/born by this technique.

Lynn (ibid) considers the extent to which AID has been eugenic 'depends on the quality of the donors' and the number of women using the method. However, he does not explain exactly what he means by 'good quality'. In the 1990s Brennan and Syal (1997) make clear, it was discovered that in Britain a number of AID clinics were paying unemployed men for sperm dominations and that 11 per cent out of the estimated 9,000 donors were long-term unemployed drug abusers, and that these men supplemented their incomes from welfare by donating semen several times a week. Lynn (ibid) finds the 'only legal restriction' placed on the use of donor sperm by fertility clinics in Britain, which is that it must first be screened for the HIV virus, 'clearly unsatisfactory'. He recommends that fertility clinics should be legally bound to screen semen donors for health and good educational qualifications.

Regarding the numbers of babies born by AID, the procedure is highly confidential with no official figures kept, so no precise details are available. Nevertheless, Caldwell (1984), estimated that around 2,000 babies a year were born in the United States in the early 1930s, through this method. By the 1970s, Feingold (1976) shows that the figure had risen to somewhere between 10,000 and 20,000. As Lynn (ibid) points out, in 1987, the United States government carried out a survey 'of the extent of the use of AID and estimated that about 33,000 AID babies were born through this means in 1986 (United States Office of Technology Assessment, 1987)'. This represents about one per cent of the total annual numbers of births in the United States. In Britain, the Galton Institute (1999) estimated that about 2,000 babies a year were born through AID during the decade of the 1990s, representing about 0.3 per cent of all births. Lynn (ibid) suggests that, 'these numbers are so low as a percentage of births that even assuming that donors of good quality have normally been used, the eugenic impact will have been negligible'. Again, the author does not explicitly say what he means by 'good quality' donors.

Egg Donation

Egg donation is the method by which a fertile woman donates her egg to a woman who is infertile, 'the counterpart of AID', as Lynn (ibid: 247) suggests. In usual conditions, the egg will be fertilised in vitro with the sperm of the husband/partner, and the embryo implanted into the uterus. Conversely, it can be implanted into another female who acts as a surrogate, bringing the pregnancy to term and handing over the baby after birth. The women who act as surrogates are usually paid a fee for the services, as are women who donate eggs. Lynn (ibid) refers to the first reported case of egg donation, which was carried out in Australia in 1984. Cohen (1996) has described the use of egg donation in a variety of economically developed countries.

This method can indeed be used for eugenic purposes if the eggs are taken from women with 'desirable' qualities that are likely to have some genetic basis. In the 1990s, a number of infertile women seeking eggs and desiring 'to obtain eggs of good genetic quality' (Lynn, ibid) advertised for them in the student newspapers of the American Ivy League colleges, including Harvard, Yale and Stanford. Lynn claims that the sum offered for eggs was $5,000, but 'sometimes as much as $50,000 has been offered'. Frequently, the advertisements specify that the donors should have good college grades, and 'certain physical characteristics, such as blue eyes' (ibid). This practice of infertile women seeking eggs from elite universities has spread to Britain, and in 1999 an advertisement for egg donors appeared in the magazine of Cambridge University, *Cam.*

Lynn (ibid) argues that such advertisements could 'no doubt obtain eggs more cheaply by advertising in the local press or by getting them from women on welfare'. However, it would appear that these women believe that they are likely to obtain eggs of, 'better genetic quality from students at elite universities' (ibid). However, this practice has not yet become widespread; and elite egg donation has not been sufficiently employed to have 'any significant eugenic impact' (ibid).

Prenatal Diagnosis of Genetic Disorders

The prenatal diagnosis of genetic disorders/diseases consists of tests 'for the genetic characteristics of the fetus carried by a pregnant woman with the object of terminating the pregnancy' in cases where a perceived genetic disorder is identified (Lynn, ibid: 248). The first testing method to be developed was amniocentesis, consisting of inserting a needle through the pregnant female's abdomen, into the amniotic fluid which surrounds the fetus, and the withdrawing of a fluid sample. Then, analysis of the sample takes place to ascertain whether or not 'defective genes' and other anomalies are present. As Lynn (ibid) shows, amniocentesis was first developed in the 1930s for the 'diagnosis of erythroblastosis fetalis, a blood disorder'. In the mid 1950s, it was utilised in Denmark to determine the sex of the fetus, and to offer termination to pregnant women who were at risk of being carriers of hemophilia. As Lynn (ibid) goes on to explain, because hemophilia is linked to an X chromosome, recessive gene disorder carried by females and inherited by about half of their sons, but rarely by daughters, it is possible to avert the risk of having a child with the disorder by ensuring that it is female. In 1967, it became possible to use amniocentesis for the fetal diagnosis of Down's syndrome; and in 1968 the first abortion of a Down's fetus was reported. By 1980, amniocentesis was being used throughout 'virtually the whole economically developed world' to test pregnant women for the presence of genetic disorders/diseases in the fetus; pregnancy terminations were offered and usually accepted in the cases where diagnosis uncovered disorder. Later, after use in Down's cases, the technique was extended to test for a variety of genetic diseases/disorders, including single-gene diseases such as Tay Sachs disease, Huntington's disease, cystic fibrosis, sickle-cell anemia, hemophilia and galactosaemia, a rare disease causing liver disorder, mental retardation and multifactorial disorders, including spina bifida etc. In the 1980s, four further techniques were developed for the diagnosis of the presence of genetic defects in the fetus. These are ultrasound scans, maternal serum screening, fetal biopsy and chorion sampling.

Pregnancy Termination of 'Defective' Fetuses

By the late 1960s, it was possible to test pregnant females for the presence of 'common fetal defects' (Lynn, ibid: 249). This lead to the development of what has become known as 'therapeutic abortion' to terminate the pregnancies where genetic defects had been discovered. The termination of such pregnancies required changes in the law to allow 'therapeutic abortion' to be carried out. Abortion was illegal in most Western countries up to the mid-1960s, except in special circumstances (such as where pregnancy had resulted from rape). As the author shows, from the mid-1960s onwards, there was a liberalisaton of abortion law, which swept through almost all the economically developed world. In Britain, the abortion of genetically-impaired fetuses, in circumstances, where the woman's health was seen to be at risk was legalised in 1967. In the United States, 'several states legalised abortion in the early 1970s, and the legality of abortion was established nationwide in1973 by the Supreme Court decision in *Roe v. Wade'* (Lynn, ibid: 250). By the decade of the 1980s, the abortion of fetuses carrying 'defects' had become legalised throughout the nations of the West, with a few examples, namely the Republic of Ireland.

Hamerton, Evans and Stranc (1993) in a Canadian study show that the percentage of pregnant women with impaired fetuses who were willing to have the fetus aborted was typically 70 per cent to 85 per cent. The proportion of women, in general conditions, opting for termination depends upon the severity of the diagnosed condition, according to the authors. For Down's syndrome, 83 per cent chose termination, and for the 'less severe' disorder of Turner's syndrome, which entails a degree of retardation and infertility in girls, 70 per cent chose termination. For 'less severe' disorders caused by added X or Y chromosomes that impair intelligence by only a relatively small amount, 30 per cent chose termination. Lynn (ibid) shows how similar figures (Royal College of Physicians, 1989) exist for other Western countries. The Canadian Royal Commission on New Reproductive Technologies (1993) found, in an opinion poll, that 75 per cent to 85 per cent of the population approve of abortions of 'severely impaired' fetuses. Weatherall (1991) showed that (in the last 1980s) 83 per cent of women aged 37 to 44 said that they would have prenatal tests for Down's syndrome and cystic fibrosis, and would have an abortion if medically recommended.

The Eugenic Implications of Prenatal Diagnosis

Owen (2006b: 912), dealing with the general eugenic potential of human biotechnology, makes the following point about eugenic practices in the modern context....

> Arguably, as Ridley (1999) suggests, what is 'wrong' with eugenics is not the science, but the coercion involved. Kitcher (1996) makes a case for calling genetic screening *'laissez-faire'* eugenics: everyone can be his or her own eugenicist, taking advantage of the available genetic tests to make the reproductive decisions he/she thinks correct. By this standard, 'eugenics happens every day in hospitals all over the world' (Ridley, ibid: 298).

We will keep this idea of Kitcher's (ibid) 'laissez-faire' eugenics (as opposed to the state eugenic programmes of Hitler's Third Reich) in mind, and return to it later in the Chapter in relation to human biotechnology.

By the 1990s, the majority of pregnant women in the so-called 'developed' nations were being given ultrasound scan and material serum screening for the detection of fetal disorders. As Lynn (ibid) shows, the impact (eugenic) of these screening procedures has been mixed. The results of a survey by the Office of Population, Censuses and Surveys (1993) into the effectiveness of prenatal diagnosis and pregnancy terminations of impaired fetuses on the birth incidence of babies for some of the most common genetic and congenital disorders reveals that there were, 'large reductions in the birth incidences of anencephaly, spina bifida, and hydrocephalus. This is because these defects are reliably detected by ultrasound scan, 'which is given routinely to 80 per cent to 90 per cent of pregnant women' attending physicians/clinics (Lynn, ibid: 251). In contrast to the reductions of such genetic disorders, there was hardly any reduction in the birth incidence of cardiovascular disorders, examphalos, and Down's syndrome. The cardiovascular disorders were reduced by only around 8 per cent, while exomphalos (a disorder in which the intestines protrude through the abdominal wall) and Down's syndrome were not reduced at all. Lynn (ibid) argues that it is evident that these disorders were not being reliably picked up by testing with ultrasound scan and maternal serum screening. With regard to the single-gene disorders, the survey reveals some 'successes' in reducing birth incidences. For Lynn (ibid), the 'greatest achievement' and hence 'positive' eugenic impact, has been, 'the reduction in the birth incidence of Tay Sachs disease among American Ashkenazi Jews to around 3 per cent of its former incidence'. Lynn is keen to emphasise the 'positive' benefits of also reducing the birth incidence of B-thalassernia, a form of anemia, among southern European populations as revealed in the work of Cantor (1992). Overall, however, as Cantor's work shows, prenatal diagnosis has resulted in a reduction of less than 5 per cent of the birth incidence of genetic diseases/disorders.

Embryo Selection

Embryo selection, as Lynn (ibid: 252) suggests, consists of growing human embryos in vitro and testing them for 'genetic characteristics, and selecting for implantation those with genetic characteristics regarded as desirable', while discarding those with 'undesirable' qualities. The procedure, also known as *embryo biopsy*, entails growing several blastocysts (embryos grown in vitro to 8 cells), removing one of the 8 cells and testing for defects of a genetic/chromosomal nature. Verlinsky, Pergament and Strom (1990) reported the use of the procedure to 'screen out' embryos and genes for Duchenne's muscular dystrophy and Down's syndrome, in order that an embryo free of such 'defects' could be implanted in the mother. At around the same time, another use for the procedure was reported by Handyside and colleagues at London University (ibid). They utlised IVF (in vitro fertilisation) for two couples in which the female partner carried an X-linked recessive disease or disorder, which can only be 'expressed' in males. In order to avoid the birth of a boy with the X-linked disorder, the physicians tested for the sex of the embryos, implanting only females. According to Lynn (ibid), 'this technique allows couples to choose the sex of their babies, whether this is to avoid having babies likely to inherit serious disorders, or simply because they prefer one sex rather than the other'. Lynn's comments appear to echo Kitcher's (1996) claim that in an age of '*laissez faire*' eugenics, couples may avail themselves of what are *eugenic* choices.

As Lynn (ibid) shows, by the 1990s there was, 'rapid progress in preimplantation diagnosis and the screening out of embryos with genetic diseases/disorders'. By 1995, the preimplantation diagnosis of embryos for genetic diseases was being carried out in 16 centres in various countries. By the late 1990s, this extended to screening for the presence of cancer genes. The first use of the method for diagnosing cancer genes was carried out in Britain in 1996, 'on the embryo of a woman with familial adenomatous polyosis (FAP), a form of bowel cancer caused by a dominant gene' (ibid). Now, it is possible to screen embryos for several thousands of genetic disorders/diseases. For Lynn preimplantation diagnosis and embryo selection is preferable to prenatal diagnosis and the abortion of 'defective' fetuses as a means of 'securing' a healthy baby. According to the author, who stresses the 'positive side' of eugenics, 'it avoids the stress of abortion, and it greatly increases the probability of having a child free of genetic diseases' (ibid). Although the procedures of preimplantation diagnosis and embryo selection 'have not been used extensively by the beginning of the twenty-first century, they are a significant eugenic advance' (ibid).

Genetic Engineering

The above term is employed here in the sense of the implantation of 'new' genes into an organism. As Lynn (ibid: 253) suggests, in the last two decades of the 20th century, new genes have been implanted into a number of plants, with success, in order to produce greater yields and resistance to diseases, and these have become known as genetically-modified food. New genes have also been implanted in a variety of animals. For example, in the early 1980s, Ezzell (1987) inserted a new gene for susceptibility to cancer into a mouse in order to research the causes of cancer. This produced the "oncomouse", the first genetically engineered new animal for which a patent was granted. The Oncomouse ™ is referred to in the previous section on the Human Genome Project, where Bucchi (2004) and Haraway (1997) make the point that it represents a 'new' technoscience as a whole, in which transgenic animals perform a similar role to transuranic elements in the days of the Cold War.

By the decade of the 1990s, as Lynn (ibid) suggests, new genes were implanted into a wide variety of animals (sheep, pigs, cattle etc) known as *transgenic-animals*. By the end of the 20th century, thousands of such animals had been produced. Lynn identifies the objectives of this work. Firstly, there is the objective or producing animals such as fish, chickens, cows and sheep that will grow more quickly, and involves the insertion of growth genes. Secondly, we have the insertion of human genes into animals in order to get them to produce organs that can be utilised for transplants into human beings. A further use of genetically engineered animals is to introduce the genes for human diseases into animals to make possible the study of such diseases. By the end of the 20th century, human genes for sickle-cell anemia, muscular dystrophy, Alzheimer's disease and a variety of cancers had been successfully implanted into mice. Finally, transgenic goats and cows have been produced in order to secrete human proteins into milk, which is then used for the treatment of human diseases such as emphysema.

Cloning

Cloning involves the reproduction of a genetically identical copy of a plant or animal. As Lynn makes clear, there is nothing 'new' about cloning in relation to plants. Plants may 'easily be reproduced by cloning by taking a cutting and putting it in the soil, and this method of 'producing identical copies' of plants, 'was known in classical Greece' (ibid: 254-255). The word *clone* is actually derived from the classical Greek word *klon*, meaning "twig".

The latter half of the 20th century saw a number of animals cloned. In the 1950s a frog was produced by the method of cloning by King and Briggs (1956), and as Lynn (ibid: 255) shows, in 1997 the first mammal, a sheep named Dolly was produced by cloning by Wilmut,

Schnieke, McWhis, Kind and Campbell (1997) at Edinburgh's Roslin Centre. Very shortly afterwards, Tranmon and colleagues at Monash University, Australia produced nearly 500 identical cattle embryos by cloning. In 1998 and 1999, a number of mice were also produced by Wakayama, Perry, Zuccotti, Johnson and Yanagimachi (1998) at the University of Hawaii. In 2003, claims by the doctor Panyiotis Zavos that he would clone a baby for less than £100,000 were dismissed as fantasy by some of Europe's leading fertility experts. The American researcher said that he was about to implant the first cloned embryo in the womb of a surrogate mother, hoping to make the procedure, 'an affordable alternative to IVF' (Derbyshire and Highfield, 2003: 8). Fertility expert Prof Ian Wilmut of the Roslin Institute, Edinburgh, who created Dolly the cloned sheep referred to above, said that he strongly endorsed a recent initiative for a worldwide ban. Wilmut was amongst a group of scientists, doctors and lawyers urging the United Nations to seek an opinion from the World Court declaring human cloning illegal. Another leading fertility expert, Prof Hans Evers, the chairman of the European Society of Human Reproduction and Embryology backed the calls for a worldwide ban. Quoted in Derbyshire and Highfield's *Daily Telegraph* article, Evers said, 'All the scientific evidence so far indicates that it is completely impossible to clone humans. Anyone claiming that he is going to clone a human being does not know what he is talking about'. The scientists and lawyers calling for human cloning to be banned, and made a crime against humanity said the technology concerned, 'all humanity' and that 'reproductive cloning represents a technique that could be used in attempts to alter human evolution, directing it towards a new species of human being' (ibid). The group also claimed that, 'the manufacture of human clones could threaten to alter the social order which humans have developed through the ages', and voiced concerns about the involvement of children as, 'experimental objects' (ibid).

The technique of cloning to create copies of human babies remains outlawed in Britain but therapeutic cloning is legal. On the 12 August 2004, the press reported that British scientists had received permission to clone human embryos for medical research, which is thought to be the first such license given in Europe. On the front page of *The Daily Telegraph*, for example, the paper's Science Editor, Roger Highfield wrote of how an attempt by researchers at Newcastle University could place the country, at the forefront of worldwide efforts to create a revolutionary generation of medical treatments. Scientists such as Dr Ian Wilmut, who cloned Dolly the sheep, 'welcomed the news as the first step in an effort to provide new insights into illness and ways to grow a patient's own cells to treat a vast range of diseases' (ibid). However, the decision was condemned by 'pro-life' campaigners as, 'tragic and frightening'. They saw it as marking the manipulation, exploitation and trivialisation of human life.

According to Lynn (ibid: 255), there are three ways of cloning mammals that could be adapted for humans. One is to remove a cell from a mature animal, revert it 'to the undifferentiated embryo state before differentiation into cells for different bodily components' (ibid), and implant it in the host mother. A second technique is to 'remove the nuclei of the cells of an early embryo or unfertilised egg' and to 'substitute the nuclei of cells taken from another embryo or adult' (ibid). A third technique would be to 'grow embryos in vitro and to split them into two or more embryos' (ibid). According to the author, this is what 'happens naturally' in the production of identical twins, so one twin, in Lynn's (ibid) view, 'can be regarded as a clone of the other'. It would be possible to produce a number of clones in this way (ibid), and the technique has been pioneered by Hall and Stillman (1993) at the

George Washington Medical Center. One of the advantages, claims Lynn, is that the genetic characteristics of one embryo can be assessed via biopsy, and if the embryo is found 'satisfactory' (ibid), the other undamaged embryo may be implanted. The technique also makes possible an increase in the number of embryos available for implantation in the treatment of infertility. Lynn (ibid) conjectures that, once the technique is perfected it will be 'feasible to use the technique to clone humans', commenting that, 'it was reported that a human embryo was successfully cloned in South Korea on December 15, 1998, and was allowed to develop into four cells, after which it was destroyed' (The Times (London), June 17, 1999).

Human Biotechnology: A Form of 'New' Eugenics?

Lynn (ibid: 256) points out that it is frequently asserted by the medical profession that human biotechnologies, 'are not eugenic because the physicians carrying out the procedures are solely concerned with the well-being of their patients' and have no ulterior motive of 'improving the genetic quality' of the population. As the author points out, by the last two decades of the 20th century, any procedure that could be identified as eugenic was automatically condemned, which may play a part in the denials from the medical profession.

Despite the 'official' denials of the medical establishment, a number of commentators have argued that the applications of techniques of human biotechnology should be recognised as eugenic because their impact affects the whole of society, rather than just the patients. Recently, the work of David and Kirkhope (2005) has explored controversies over human cloning, specifically the media's role in escalating public fear around such techniques by calling on 'scare stories' from Aldus Huxley's *Brave New World* and Ian Levin's *The Boys from Brazil.* The authors attempt to address the current impasse between technically optimistic market individualism and traditional (often religious) forms of moral collectivism (ibid). However, those who express suspicions about the eugenic nature of the medical application of eugenics are considerable in number. For example, Duster (1990) argued in *Backdoor to Eugenics* that such techniques represent a 'covert' reintroduction of eugenics. Lippman (1991) suggests that it is hypocritical to deny the eugenic nature of the application of human biotechnology because....

> Though the word eugenics is scrupulously avoided in most biomedical reports about prenatal diagnosis, except where it is strongly disclaimed as a motive for intervention, this is disingenuous. Prenatal diagnosis presupposes that certain fetal conditions are intrinsically undesirable (p. 24).

According to Lynn (ibid: 256) the issue of whether the biotechnologies, such as those previously listed under sub-heading in this Chapter, are 'eugenic' depends *either* on the intentions of 'the physicians carrying out the procedure' or 'by their effects'. Owen (2006b: 912) has argued that Lynn is wrong here. Arguably, 'to analyse the issue in terms of the *effects* of the procedures would be to engage in *functional teleology*, an illegitimate form of reasoning' and one of the 'cardinal sins' of the metatheoretical framework employed here (ibid: 28). Here, we must, as Owen (ibid) suggests, 'regard the physicians as *actors*, in other words, entities that are capable of making, taking and acting upon decisions'. Lynn (ibid) maintains that the criteria of the intentions of the physicians carrying out the procedures is 'unsatisfactory' because it is difficult to know what their intentions are, 'especially since the physicians are going to admit publicly that they are endeavouring to promote eugenics'. Anonymous surveys have revealed a mixed result; some physicians *do* consider that they are practicing eugenics, while others do not. There is some evidence that appreciable numbers of professionals carrying out prenatal diagnosis and counseling patients on whether to terminate fetuses with genetic diseases/disorders, are in fact motivated by eugenic considerations concerning the genetic health of the population and not solely by the well-being of individual patients. A large survey by Wertz (1998) was conducted in 1995 in 36 nations, and surveyed the views of 2,901 genetics professionals and physicians. In the survey, as Lynn (ibid) shows, many of them advised their patients to have terminations and agreed with the statement, "An important goal of genetic counseling is to reduce the number of deleterious genes in the population". As Lynn (ibid) suggests, 'this is certainly a eugenic objective'. All of the 252 professionals in China agreed with the statement. Many also agreed with the statement, 'It is socially irresponsible knowingly to bring an infant with a serious genetic disorder into the world in an era of prenatal diagnosis'. In the United States, 55 per cent of primary care physicians and 26 per cent of geneticists, in addition to 44 per cent of patients, agreed.

In what follows, we consider more deeply the ethical/eugenic concerns pertaining to selected examples of human biotechnology, having covered the developments of the six major techniques as identified by Lynn (2001). Afterwards, we look finally at the *'Bio-medical model'* (Powell and Owen, 2005) in order to emphasise the need to avoid accounts which rely upon biological *reductionism* as well as the *oversocialised* conceptions of the person found in Gagnon and Simon's work (1973) when developing a metatheoretical framework for, in this case, the study of related examples of human biotechnology. Some terms from the framework are applied to a critique of the Bio-medical model, but not in the depth that we shall encounter in Chapter Five, which is specifically devoted to the application of the framework to the subject. The developments in biotechnology previously detailed have raised concerns over whether they are ethically acceptable and permissible. We have looked briefly at the eugenic potential of selected examples. Now, it is our task to explore ethical issues in greater depth in relation to the developments in human biotechnology. Here, we are to examine issues pertaining to AID; egg donation; prenatal diagnosis; embryo selection; selection for sex; cloning; gene therapy and the case for permitting germ-line gene therapy.

FURTHER ETHICAL CONCERNS AND HUMAN BIOTECHNOLOGY

Human biotechnology, as Lynn (ibid: 258) points out, has given birth to a new academic discipline – *bioethics*, with its own journals in which ethical question are addressed and debated. It is *not* the intention here to cover the literature on *bioethics* per se. That is beyond the scope and intention of this book. However, it is necessary to cover the major ethical concerns pertaining to selected examples of human biotechnology. One could not credibly cover the subject of biotechnology without discussing ethical concerns because the two are linked so strongly in the public perception (David and Kirkhope, 2005; Wilkie, 1994). As Lynn (ibid) points out, although techniques of human biotechnology have a 'eugenic impact', the ethical problems are 'not the same as those raised by classical eugenics'. As the author shows, in classical eugenics, the state (which, in the metatheoretical framework is *not* regarded as an actor because it does not posses the *agency* to formulate and act upon decisions) can legitimately interfere in the population's reproduction. An example of this would be the eugenic measures carried out by the state in Hitler's Third Reich. In Western democracies, the state does not promote the use of human biotechnology to influence the population's fertility and promote eugenics. As Lynn suggests, the biotechnologies are used by *individuals* to enable them to have children and 'by most to have children with genetically desirable qualities' (ibid). Here, Kitcher's (1996) suggestion that we are living in an age of '*laissez-faire*' eugenics appears to apply. On the surface, this sounds as if everyone can be their own eugenicist, taking advantage of available genetic tests in order to make the reproductive decisions she or he views as 'correct'. However, Ridley (1999: 298), asks an important question, in the light of Kitcher's *'laissez-faire'* eugenics....

> Have we rejected government eugenics merely to fall into the trap of allowing private eugenics? Parents may come under all sorts of pressures to adopt voluntary eugenics, from doctors, from health-insurance companies and from the culture at large. Stories abound of women as late as the 1970s being cajoled by their doctors into sterilisation because they carried a gene for a genetic disease. Yet if government were to ban genetic screening on the grounds that it might be abused, it would risk increasing the load of suffering in the world: it would be just as cruel to outlaw screening as to make it compulsory. It is an individual decision, not one that can be left to technocrats.

Lynn (ibid: 259) makes the point that, 'in liberal democracies the citizens should be permitted to act as they please unless there is a reasonably strong case that by doing so they cause harm to individuals or inflict social damage'. That, the author maintains, is the very basis for the protection of individual freedom in liberal democracies as formulated by political philosophers such as Mill (1859) and Hayek (1960). To justify the prohibition of human biotechnologies, 'it has to be shown' that the techniques, 'are likely to cause harm either to the individuals born through their use or to society as a whole' (ibid). There is support for this standpoint in the work of Peters and Bennett (2004) and Kerr (2003).

Artificial Insemination

This technique raises two major ethical issues. As Lynn (ibid) suggests, firstly, there is the question of whether it is acceptable to use AID at all, and the second is whether it is acceptable to select donors 'for what are likely to be desirable genetic qualities'. The Roman Catholic Church condemns AID on the grounds that 'the only ethically acceptable way for women to have children is through sexual intercourse' (ibid). Ratzinger and Bovone (1987), explained the Catholic view in a Vatican publication: 'Artificial fertilisation violates the right of the child: deprives him of his filial relationship with his parental origins and can hinder the nurturing of his personal identity' (p.23). Lynn (ibid) makes the counter point that 'no persuasive evidence has been produced' to show that children born through such techniques suffer psychological impairment, and that, 'it is a curiosity of Catholic doctrine condemning AID that Jesus Christ and his mother, Mary were both, according to Catholic doctrine, conceived by AID'.

The second ethical issue in the use of AID centres around the selection of donors judged to have desirable qualities. When women make the decision to have babies via the technique of AID, they and the physicians involved face the dilemma of selecting the donors who supply the sperm. The donors may be, 'and in practice usually are' (ibid) selected on the grounds that they have 'good qualities' regarding 'health, intelligence and personality', that such traits have a likely 'genetic basis' and will be transmitted to the offspring. As Lynn points out, for such reasons it is 'rare' that women utilising this procedure and their physicians 'use the semen of those with serious genetic illnesses, the mentally retarded, or criminal psychopaths' (ibid). Lynn shows how, when Robert Graham established the Repository for Germinal Choice in 1980 as a sperm bank for the storage of Nobel prizewinners, a number of critics, such as the bioethicist Arthur Caplan objected that it was 'morally pernicious'. However, as the work of Bojorquez (1994: 18) shows, Caplan changed his view and concluded, 'we mold and shape our children according to environmental factors. We give them piano lessons and every other type of lesson imaginable. I'm not sure there is anything wrong with using genetics as long as it is not hurting anyone'. Amongst those non-Catholics who find AID unnatural and ethically abhorrent were two British members of parliament in 1991; Jill Knight and Jerry Hayes. According to the *Washington Post* (March 12, 1991, p.8), Knight asserted "It is difficult to imagine a more irresponsible act than to assist a woman to have a child in this highly unnatural way", and Hayes stated, "I find it personally abhorrent".

Egg Donation

Lynn (ibid: 261) shows how in Britain, the government in 1990 sought in its Human Fertilisation and Embryology Act to 'restrict egg donations' by limiting the fee payable to egg donors to £15 plus minimal expenses. In Lynn's view, this is unreasonably low and, 'invites evasion by the surreptitious payment of higher fees' (ibid). For the author, in a 'free society', there can be 'no legitimate ethical objections to infertile women buying eggs from whoever they please and paying whatever price for them is mutually acceptable to willing buyers and sellers'. On this issue of payment for egg donation, there have been recent developments. As Boseley (2005: 4) points out, the government have recently thrown the debate over the future of fertility treatment open to the general public, asking for views on a range of controversial issues, and payment for those who donate their eggs is a key issue. The British government wanted to examine public views before it legislated to modernise the Human Fertility Act, in the words of Boseley (ibid), 'probably sometime between now and 2008'. The payment of £15 plus expenses is being reviewed in the light of the removal of anonymity from donors, which is severely reducing the numbers coming forward. Boseley's position here, regarding governmental motives appears to differ from that of Lynn (ibid), who sees government as 'restricting' egg donation. The forthcoming European Union directive says donors should be unpaid although they can receive compensation for expenses and inconvenience. The paper, according to Boseley (ibid) asks whether it should be the Human Fertilisation and Embryology Authority which decides how much donors receive, or whether that role should pass to parliament. As of the time of writing, British MPs have voted overwhelmingly in favour of radical stem cell research which would enable scientists to create human-animal embryos [Daily Telegraph, 20th May 2008]. MPs also voted against an amendment to the

Human Fertilisation and Embryology Bill to ban the creation of saviour siblings, where parents use fertility treatment to select an embryo that is a genetic match of an older child who has a disease. The present Prime Minister Gordon Brown and Conservative leader David Cameron, who both have sons with severe medical conditions, united to support the creation of the hybrids, known as ''admixed embryos''. The move was opposed by religious leaders, including Cardinal Keith O'Brien, the leader of the Roman Catholic Church in Scotland.

Prenatal Diagnosis

This technique raises three major ethical issues: firstly, whether the technique is ethical in itself; second, whether it is ethical to have prenatal tests and terminations for non-serious disorders or for conditions that are not disorders such as the sex of the fetus; and thirdly; whether it is ethical *not* to have the tests and terminations (Lynn, ibid: 262). The author makes the point that the major ethical justification for prenatal diagnosis and the termination of pregnancies in cases where a genetic disorder is present, 'lies in the general principle that in a free society individuals are permitted to make their own decisions concerning the conduct of their lives', unless there is a reasonable case that their behaviour may occasion social harm (ibid). However, again, the Roman Catholic Church and certain other religions reject its ethical legitimacy on the grounds that all abortion is ethically unacceptable. For Lynn (ibid), the abortion of a, 'genetically disordered fetus cannot be regarded as likely to cause social harm' and hence can be ethically justified.

The second ethical issue raised in relation to prenatal diagnosis concerns the abortion of fetuses with non-serious disorders. As Lynn (ibid) points out, the problem in attempting to differentiate between serious genetic disorders for which it *may* be seen as ethical, and between non-serious disorders for which it *may* be seen as unethical to have an abortion, is that there are more than 6,000 genetic diseases/disorders in a spectrum from those that are 'totally disabling' through to 'less serious disorders (such as Down's syndrome) to those that are mild disorders (such as dwarfism)'. The author argues that it would not be feasible to reach any agreed consensus on a classification of these cases into the disorders that are 'sufficiently serious to allow prenatal diagnosis and termination, and those that are insufficiently serious and for which this procedure would be prohibited' (ibid: 262-263).

The third ethical issue raised by the procedure of prenatal diagnosis is whether it is ethical for women to refuse the procedure. As far as Lynn (ibid: 263) is concerned, 'where the genetic disorder is serious, there is a strong case that such women are acting unethically because they are bringing into the world a child whose health will be seriously impaired, causing distress to itself and to its family, and whose maintenance imposes significant costs on society' (ibid). Lynn applies this position to children with Down's syndrome. To incur the 'risk of giving birth to such a child by refusing to have a prenatal diagnosis' can 'reasonably be argued as unethical'. This is the view taken by the American Society of Human Genetics, which has asked the question of "whether or not a defective fetus should be allowed to be born", and suggested that the ethical answer is in the negative (Shaw, 1984: 1). Lynn (ibid) shows that Robert Edwards, a British physician who practices prenatal diagnosis and the termination of pregnancies, argued in 1999 that, 'soon it will be a sin of parents to have a child that carries the heavy burden of genetic disease'. Edwards urged that all pregnant women should be tested for Down's syndrome and common genetic disorders, and that it would be 'unethical for women to refuse to have the tests carried out and to refuse to have fetuses with serious disorders terminated' (Rogers, 1999: 28).

Paul (2000) suggests that attitude surveys indicate that opinions about prenatal testing are, 'both favourable and confused' (p.264). For example, about two thirds of the respondents to one survey said that they would want to undergo such tests (or would want their partner to); they believe that tests will do more good than harm. However, they understand very little about them. Recent work by Sturgis, Cooper and Fife-Schaw (2005) supports the view that public 'scientific literacy' is generally low, falling well short of what normative criteria would consider 'satisfactory'. Regarding Singer's (1991) respondents, a majority thought that current tests could predict whether or not a person will suffer a heart attack and that genetic screening could correct the genetic defects that cause such inherited diseases as Down's syndrome. Tay Sachs disease and hemophilia.

Embryo Selection

As Lynn (ibid: 264) suggests, one ethical position on embryo selection is that it is 'unacceptable' in every circumstance. This is the position adopted in several European countries and US states, which in the 1990s decade, prohibited all uses of embryo selection. These prohibitions were introduced in Austria, Germany, Norway and Spain and in the American states of Minnesota, New Hampshire, Louisiana and Pennsylvania. Such prohibitions, in Lynn's (ibid) view, 'cannot be justified in these countries and states that allow the abortion of fetuses with genetic diseases and disorders but prohibit the rejection for implantation of embryos' with the very same genetic diseases/disorders. For Lynn, embryo selection is more ethically acceptable than abortion because the procedure takes place earlier or when most of the embryos that are grown in vitro have been destroyed.

A principal argument against negative embryo selection is that it may stigmatise those who have the disorders for which the embryos are originally screened and rejected. A committee of the European Parliament (1990: 29) stated that the use of embryo selection, 'undermines our ability to accept the disabled'. Testart (1995) takes a very similar position. For Lynn (ibid), the argument 'should apply equally strongly to the prohibition of prenatal diagnosis' because 'both procedures are based on the assumption that it is preferable for babies to be born healthy' than with genetic disorders. Lynn claims that the argument 'cannot be accepted' (ibid) because the gains of having healthy babies must be greater than the psychological distress experienced by those suffering from the genetic diseases. Unfortunately, the author does not appear to offer any evidence for this claim.

Several who have examined ethical problems pertaining to embryo selection draw a distinction between 'negative' and 'positive' embryo selection. For example, the European Parliament in 1990 declared that genetic techniques and practices, 'must on no account be used for the scientifically and politically unacceptable purpose of 'positively' improving the population's gene pool', and called for, 'an absolute ban on all experiments designed to reorganise on an arbitrary basis the genetic make-up of humans' (Kevles and Hood, 1992: 320). Kevles and Hood (ibid) commented that 'the idea that genetic knowledge will soon permit the engineering of Einsteins or even the enhancement of general intelligence is simply preposterous'. Bodmer and McKie (1994) make similar distinctions between the 'permissible' rejection of genetically-disordered embryos, and the 'unethical' selection of embryos for intelligence and personality.

As Lynn (ibid: 265) suggests, some of those who have accepted the ethical legitimacy of 'negative' selection but reject the legitimacy of 'positive' selection for intelligence, personality, appearance etc, have voiced anxieties that even though 'negative' embryo selection is acceptable it will lead to the eventual use of 'positive' embryo selection. In 1995, as Lynn (ibid) shows, John Polkinghorne (a professor of physics at Cambridge University and a Member of the Church of England General Synod) 'conceded the ethical legitimacy of negative embryo selection' but expressed great anxiety at this being the first step towards the more broader use of embryo screening. Lynn regards this is as a typical 'slippery slope' argument to the effect that although negative embryo selection is legitimate in ethical terms, it could lead to 'unacceptable' positive embryo selection. Lynn (ibid) argues against such a position by stating that it should be possible to permit negative embryo selection by drawing up a list of the genetic disorders/diseases that could 'lawfully be diagnosed and rejected', whilst at the same time 'prohibiting positive embryo selection'. An argument, sometimes advanced against positive embryo selection for intelligence and other qualities is that the procedure would be expensive and, could only be afforded by the wealthy, as Roberts (1996) suggests. Roberts (ibid) has brought a radical dimension to the argument by suggesting that whites would be able to use embryo selection more than blacks, on grounds of greater relative affluence. This, the author suggests, is unjust and is an argument for prohibiting both IVF and embryo selection. As Owen (2006b:913) has pointed out, Nelkin's (2001) work advances a similar argument against the commodification of genetic products, and conjures up, 'hypothetical scenarios of 'war' between 'the gene rich' and the 'gene poor'.

Selection for Sex

The selection of a child's sex is a development of human biotechnology which has aroused some ethical and critical concern. Lynn (ibid) points out that there are three techniques by which this can be achieved. Firstly, there is the staining of a sample of semen with dye. Sperm containing two X chromosomes, (which develop into females) are heavier and absorb more of the colouring (Lynn, ibid). Sperm of the desired sex are then given to the mother by artificial insemination. This procedure is legal in Britain and the US, and clinics were established in the 1990s, such as the IVF Genetic Center in Fairfax, Virginia, and the London Gender Clinic (ibid). Secondly, there is the practice of ascertaining the sex of the fetus and aborting it if it is of the undesired sex, allowing the couples to try again until they conceive a child of the desired sex. Thirdly, IVF can be used 'to grow several embryos, the sex of these can be ascertained, and the embryo of the desired sex can be implanted' (Lynn, ibid: 266). The last procedure was made illegal in a variety of European countries in the early 1990s, under the general prohibition of 'any use of embryo selection' (ibid). It was made illegal in Britain in 1997. However it has remained legal in Italy. For Lynn, this shows the 'ineffectiveness' of prohibiting human biotechnology in particular countries, and the 'loopholes' inevitably present when procedures are allowed in other countries.

It is widely believed that sex selection is ethically unacceptable (Lynn, ibid; Wilkie, 1993). This is the basis for the prohibition of selection for sex in several European countries. Nevertheless, as Lynn argues, there are several arguments for permitting the selection of babies for sex. As the author suggests, 'it is by no means certain that there is a widespread preference for boys in the United States and Europe or elsewhere in the economically developed world', (ibid) in contrast to the cases of China and India, where significant numbers of baby girls are killed (Lau, 1995). Pebley and Westaff (1982) found that (in the United States in the 1970s) approximately two-thirds of women would prefer a boy for their first child, however, they also found that most people would like to have two children. In Lynn's view (ibid), 'whether many of these would want two boys is open to doubt'. Even if sex selection develops 'on a significant scale' and leads to an 'excess of males', this could be 'desirable' argues Lynn (ibid), as in such a society males would have to compete more energetically for females, and in this competition males with higher status would tend to succeed. Conversely, as Buss (1999) suggests, women would have greater choice of men, and would tend to select those with higher status. (Lynn ibid: 267) sees the likelihood of 'adverse consequences' arising from allowing couples to choose the sex of their children as 'too remote a contingency for this freedom to be prohibited. Lynn argues on 'libertarian grounds' for the permission to exercise this choice. Interestingly, Murphy (2000) takes things a stage further, and asks the question of whether parents should be recognised as having the 'right' to control the sexual *orientation* of their children. Murphy's (ibid: 227) view is that parents would be within their political rights to use 'any genetic or other controls' that afforded them the possibility of control over their children's sexual orientation. This is, 'not to say that a parent has the right to a straight child' (ibid). Rather, Murphy appears to be arguing that as a matter of social policy it would be unwise to impose barriers to the use by parents of prenatal diagnostics and interventions.

Cloning

In the last two decades of the twentieth century, cloning became widely condemned (Lynn, ibid). Warnock (1987), a leading British ethical philosopher and authority in human reproductive biotechnology, condemned the practice on the grounds....

> That there must be some things which, regardless of consequences, should not be done, some barriers which should not be passed. What marks these barriers out is often a sense of outrage, if something is done; a feeling that to permit some practice would be indecent or part of the collapse of civilisation (p.8).

As Lynn (ibid) shows, the American bioethicist R.A. McCormick (1994) condemned the cloning of human beings on the grounds that it would be contrary to, 'our cherished sense of the sanctity, wholeness and individuality of human life' (p.16). The cloning of humans was legally prohibited in Britain in 1990 by the Human Fertilisation and Embryology Act. In America, the National Institute of Health set up a Human Embryo Research panel in the 1990s to 'consider the ethical issues concerning in vitro technology in general and to advise on what research should be acceptable' for federal funding (Lynn, ibid). Among the recommendations for blacklisting for funding were cloning and sex selection of embryos, except 'to prevent genetic diseases linked to the X chromosome' (ibid). However, as Lynn reveals, the panel did not recommend that these should be prohibited, rather the recommendations were that, 'it should remain legal for privately run IVF clinics to provide these services' (ibid).

In 1997, shortly after the cloning of the sheep Dolly, Lynn (ibid: 268) shows that many 'eminent persons' called for the prohibition of the cloning of human beings. Kolata (1997: 6) shows how the embryologist Ian Wilmut (who produced the cloned sheep, Dolly) said of human cloning; 'all of us would find that offensive'. Interestingly, Wilmut has recently made the news (*The Daily Telegraph,* news bulletin, December 28, 2005) by arguing that controversial stem cell therapy should be used on terminally ill patients. According to the newspaper report, Wilmut (appointed in December 2005 as the first director of Edinburgh University's new Centre for Regenerative Medicine) believes that such experimental treatment could save lives or at least speed up the pace of research. He also said that he knew terminally ill patients who would be prepared to participate in such trials. Kolata (ibid: 196) mentions that the then American President, Bill Clinton stated in June 1997 that to fund the cloning of human beings would be 'morally reprehensible'. Joseph Rotblat, a British Nobel Prize winner, described human cloning as a, 'means of mass destruction' (Harris, 1998). As Lynn (ibid) shows, the Director General of the World Health Organisation (WHO), Hiroshi Nakajima, stated, 'WHO considers the use of cloning for the replication of human individuals to be ethically unacceptable as it would violate some of the basic examples which govern medically assisted procreation' and these included 'respect for the dignity of the human being' and 'protection of the security of human genetic material' (World Health Organisation, 1997). Frederico Mayor, the Director of UNESCO, issued a statement that, 'human beings must not be cloned under any circumstances'. As Lynn (ibid) shows, a couple of days later, the European Parliament (1997) passed a resolution which condemned cloning in the following fashion.

> In the clear conviction that the cloning of human beings cannot under any circumstances be justified or tolerated by any society, because it is a serious violation of fundamental rights and is contrary to the principle of equality of human beings as it permits a eugenic and racist selection of the human race, it offends against human dignity and it requires experimentation on humans. Each individual has the right to his or her own genetic identity, and human cloning is, and must continue to be prohibited (p. 84).

According to Lynn (ibid), there are three major ethical objections that can be made to the cloning of human beings. The first concerns the point made by the European Parliament, in which it is suggested that 'the security of genetic material would be compromised'. In Lynn's view, it is rather difficult to discern the exact meaning of the phrase, and if what is intended is the 'preservation of genetic variability', the reduction of this by the production of a 'small number of clones' would be 'negligible' (ibid). It will be recalled that many, from American Presidents to genetic scientists, view cloning per se as immoral.

Secondly, according to Lynn (ibid: 268), we must consider the objection put forward in the resolution of the European Parliament, which condemns cloning for its 'eugenic' and 'racist' selection of the human species-potential. For Lynn, whilst it is 'undoubtedly true' that cloning could be put to use in order to replicate individuals 'who might do a lot of damage' (ibid), the same argument could be made against several other techniques of biotechnology, such as embryo selection and even against 'normal reproduction'. In Lynn's view, 'the cloning of undesirables is unlikely to occur on a sufficient scale in the Western democracies to justify the prohibition of all cloning' (ibid). The author does not appear to provide evidence to support his claim here. Opinions on the ethical implications of biotechnology appear to shift over time, as is suggested by the work of Parry (2003). Parry (ibid) looks at the politics of cloning; the mapping of the 'rhetorical convergence' regarding embryo and stem cells in parliamentary debates, focusing upon the rhetorical strategies and maneuverings over the meanings of stem cells, cloning etc in the debates around the April 2001 amendment of the 1990 Human Fertilisation and Embryology Act (HFEACT) which allows stem cell research to use human embryos.

The third ethical objection to human cloning, as identified by Lynn (ibid: 269) is that, 'a cloned individual would suffer a loss of personal identity'. The author argues that, 'it is contended that we all feel ourselves unique, that cloned individuals would be deprived of this sense of uniqueness', and that it would be 'psychologically damaging'. The author suggests that the problem with such an argument lies in the fact that identical twins (who are 'genetically identical in precisely the same way as clones'), do not appear to experience psychological damage from the knowledge that they have a genetically identical brother/sister. Furthermore, the author suggests that although a cloned individual would be 'genetically identical to the parent', it would not be 'psychologically identical', having been reared in 'a different environment', having quite different experiences. For Lynn, valid ethical objections to human cloning would have to show that it caused the cloned individual 'psychological damage' or would be 'likely to cause social damage' (ibid: 270). Lynn regards neither of these arguments plausible. However, it would appear to be the case that we do not yet have any evidence for Lynn's claims, and we cannot be certain that cloned individuals would not suffer psychological damage. Cloning remains prohibited in Britain except in the case of the cloning of human embryos for medical research/therapeutic purposes.

Gene Therapy

As Lynn (ibid: 270) suggests, in the 1990s, there were 'several successful trials' for the treatment of diseases by gene therapy. Here, healthy genes are inserted into patients in order to 'correct the deficiencies' of their malfunctioning genes (ibid). We need to make a distinction here between *somatic-line* and *germ-line* therapy, if we are to examine the ethical issues surrounding this form of biotechnology. The insertion of 'healthy' genes into somatic-line cells only affects the individual patients themselves, whereas germline therapy involves the insertion of genes into germ cells, which are transmitted to children, and the practice is therefore 'eugenic' (Lynn, ibid). Germline therapy has been attacked as 'unethical', and its use on human beings has been prohibited in Britain, the United States, and much of Continental Europe, 'on the grounds of possible adverse and unknown future consequences' (ibid). Suzuki and Knudtson (1990) make a case for prohibiting germline gene therapy on humans. As Lynn (ibid), shows the authors argue firstly that even apparently harmful genes may be 'useful', and that therefore it can sometimes be a mistake to eliminate them. Sickle-cell anemia is the example advanced in support of the argument. The disease is caused by inheritance of two recessive genes for the condition. The genetic inheritance of one recessive gene for the disease has no adverse effects, and actually provides immunity from malaria (Lynn, ibid). However, despite the claims of Suzuki and Knudtson (ibid), Lynn considers that this does not mean 'that there is a sound argument for preserving the gene, especially amongst populations in temperate environments where there is no malaria' (ibid). As Lynn suggests, it might be preferable to prevent malaria by immunisation and to eliminate the sickle-cell gene. Additionally, the disease itself is unusual in the sense that a single copy of the recessive gene 'confers an advantage' (ibid). There are 'no other genetic diseases known to have an advantage of this kind' (ibid). So, as far as is known, the genes for genetic diseases are 'wholly disadvantageous', and Lynn recommends their elimination by the means of germline therapy.

The second argument advanced against the use of germ-line therapy by Suzuki and Knudtson (1990), as Lynn suggests, is that once the technique was accepted it might lead to eugenic interventions for other characteristics such as intelligence. The authors (ibid: 204) have argued that, 'the history of genetics suggests that once a human characteristic – such as a particular skin colour or poor performance on IQ tests – has been labelled a genetic 'defect', we can expect voices in society to eventually call for the systematic elimination of these traits in the name of genetic hygiene'. Lynn (ibid: 271) regards this as a 'slippery slope' argument, the general format of which is that, 'an innovation may be beneficial but should not be allowed because it might lead to other innovations that would be harmful'. The author makes the point that this principle 'would prevent much and possibly all technical and scientific progress, virtually all of which has potentially harmful applications' (ibid).

Suzuki and Knudtson's (ibid) third argument against germline gene therapy is that the elimination of harmful genes would reduce genetic diversity, and that such diversity should be preserved. Their argument is that genetic diversity is 'desirable' because any genes might at some point in time prove useful in combating some unpredictable problem/hazard. Lynn's view is that, in effect, Suzuki and Knudtson are in favour of some unfortunates enduring the suffering caused by a genetic disease 'on the off chance' that at some future time their 'defective' genes might provide 'some sort of unknown benefit' (Lynn, ibid).

Fourthly, Suzuki and Knudtson (ibid: 48) have suggested that….

> While genetic manipulation of human somatic cells may lie in the realm of personal choice, tinkering with human germ cells does not. Germ-cell therapy, without the consent of all members of society, ought to be explicitly forbidden.

As far as Lynn (ibid) is concerned, the 'use of the term *tinkering* for germ-line therapy, betrays the emotive stance of the writers'. However, should *emotive* stances be considered 'wrong' or 'inappropriate? As the work of Ratzinger and Bovone (1987) suggests, issues pertaining to the use of human biotechnology *do* produce *emotive* responses from many, many human beings. For Lynn (ibid), the technique of germline gene therapy should no more be labelled *tinkering* than should heart/liver transplants provided to patients with heart disease/renal failure. Additionally, Suzuki and Knudtson's 'stipulation that every single citizen would have to consent to the use of germ-line gene therapy before it can be sanctioned, 'is a very stringent requirement' which implies that a single objector in a nation of millions should have the ability to veto legislation of the genetic procedure (Lynn, ibid).

Nelkin (2001), in the course of discussion upon the *commodification* of genetic products, argues that a new kind of inequality may emerge between those who can afford genetic services and those who cannot (a similar case was made by Roberts (1996)). Like cosmetic procedures, high cost genetic services are, in the author's view, highly unlikely to be reimbursed by public or private insurance programmes, and access to services would be limited, based upon fiscal wealth. According to Nelkin (ibid), gene-enhancement treatments could lead to the emergence of a new socio-economic class comprised of those with what are deemed to be 'superior' genetic endowments. Wealth-based access to genetic enhancement therefore creates a moral/ethical challenge, but it may also constitute a 'political threat' and a basis for social tension. As Owen (2006b: 913) points out, Nelkin's arguments against the commodification of genetic products may appear to be based upon 'feelings of *emotive aversion*', conjuring up as they do, 'hypothetical scenarios of 'war' between 'the gene rich' and 'the gene poor'. However, that in itself does not make the arguments 'wrong' or invalid. There are considerable difficulties in the regulation of gene therapy, according to Nelkin (ibid). The author draws attention to a regulating vacuum in the case of research that is supported by private industry where there are very few requirements for public disclosure and little regulation. Nor, as the author emphasises, are regulatory bodies fully equipped to deal with the social and ethical issues involved in new developments in human biotechnology. In the United States, the regulation of biotechnology focuses upon the safety and efficiency of genetic products rather than on their potential social and ethical implications. Europeans appear to have been more willing to consider the moral/social issues addressing germline gene therapy through the broad framework of international human rights law and the European Commission. As Nelkin (ibid) states, the European Commission has stated that technologies that have the potential to alter the human genome serve to compromise human dignity and are highly unacceptable.

However, just as 'performance-enhancing drugs have proliferated – available in some foreign markets – so too will offshore or 'black market' genetic enhancement services' (Nelkin, ibid: 2). The problems of regulating genetic manipulation are compounded by problems of *distinguishing between* therapeutic and enhancement uses. For example, the author (ibid) refers to a genetic treatment called *BioBypass*, used to foster growth of blood vessels in heart tissue to avoid the need of bypass surgery, and claims that it can also be utilised to grow blood vessels in *healthy* skeletal muscles to give athletics greater strength.

As Nelkin (ibid) indicates …

> Commodification of genetic products raises many questions. Will technology firms, concerned about creating a market for their products and services, influence the definition of normalcy and disease? Clearly, commercial interests have expanded the consumer demand for cosmetic services. Why not for gene enhancement? What is the difference between providing braces to straighten teeth and providing a hormone to increase height or a gene to improve memory? What is pathological or normal? Enhancement or treatment?

The author concludes by reflecting that we live in an age of invasive, aggressive commercialism, and the 'new' genetics (particularly gene therapy) has the potential to reshape our perception of children, fostering the idea of them as *commodities*. It is Nelkin's fear that children will be increasingly viewed in terms of their biological potential for success in the 'commodity culture' - a potential that can now be engineered, enhanced or manipulated by genomic techniques.

As we have seen, Lynn (2001) is sceptical about the various ethical cases against the permission of germline gene therapy. He advances a case for the permission of the genomic techniques based upon two considerations. Firstly, germline therapy 'corrects the genetic defect once and for all', whereas somatic-line therapy 'allows the defective genes to be transmitted and entails the costs, hazards and inconvenience of medical interventions for all the descendants inheriting the disorder' (ibid: 271-272). A second argument for the permission of germline therapy is that 'there are certain cases where this is the only way of allowing couples to have a healthy child' (ibid: 272). One of the situations is where, 'both parents have the same recessive gene disorder' (ibid). A further instance is 'women with genetic mitochondrial diseases, which are passed to all their children' (ibid). In both cases, as Lynn makes clear, couples are unable to have a disease-free child by prenatal diagnosis or prenatal implantation diagnosis, as all their children inherit the disease. Here, the 'defective' gene would be repaired in the ovum before fertilisation. The ovum would be removed from the woman, and the 'defective' gene replaced by a 'healthy' gene from a donor. The ovum would then be fertilised in vitro. Although, as we have seen here, germline gene therapy has been widely condemned, there are dissenters aside from Lynn (ibid), who are in favour of its permission. These include Koshland (1988), editor of the influential journal, *Science;* and Rubenstein, Thomasma, Schon and Zinaman (1995), physicians at the medical school of Loyola University in Chicago.

Lynn (ibid: 273) claims that the ethical objections raised to the selected examples of human biotechnology consists of, at least in part, 'yuk reactions' and partly of attempts to show that these genomic techniques carry the risk of harming either society in general, or the people involved. As Lynn (ibid) indicates ....

> The yuk arguments against the use of biotechnology to improve human genetic quality are unpersuasive because yuk reactions to various behaviours have changed quite rapidly over time. In previous historical periods, people had yuk reactions to those who professed atheism and to women who displayed their ankles, both of which were widely regarded as unethical. In the twentieth century there were widespread yuk reactions to the sale of contraceptives, to abortion, to artificial insemination by donors, and to in vitro fertilisation and the like. Today, the great majority of people no longer experience yuk reactions to these things and accept them as normal and ethically legitimate. We can anticipate that it will be the same with the more recent developments in human biotechnology, like embryo selection, cloning, and germ-line gene therapy, as these are developed and people become more used to them.

Perhaps as Sturgis et al (2005) suggest, public 'scientific literacy' is generally low, but it may be the case that Lynn (ibid) underestimates the depth of feeling against the use of human biotechnologies such as cloning, the genome-altering, germline therapy and fears of a return to eugenics. Lynn's view is that 'yuk' reactions do not form a secure basis for ethical judgments because they are essentially 'feelings' of emotional aversion 'unsupported by rational arguments'. Those such as Suzuki and Knudtson (1990), Ratzinger and Bovone (1987) and Warnock (1987) appear to be perfectly rational in their arguments against the eugenic potential of these techniques. As previously stated, who is to say that human *emotional* responses to certain issues such as cloning, eugenics etc are invalid and 'wrong'?

It is at this stage of the Chapter, having provided a general overview regarding the issues and implications surrounding selected examples of human biotechnology and the controversial Human Genome Project, together with a focus upon selected perspectives from the sociology of science, that we come to examine what Powell and Owen (2005) call '*the Bio-medical model'*. As previously stated at the beginning of the Chapter, whilst it is essential to avoid *oversocialised* perspectives which deny genetic causality altogether, the emphasis here is upon Ridley's (1999, 2004) conception of '*nature via nurture*'. In other words, we acknowledge that we need to look at the *biological variable* (evidence of, at least partial, genetic causality) alongside *agency-structure, micro-macro* and *time-space* when engaged in analysis, but equally we need to avoid *essentialist, reductionist* totalised discourses such as the '*Bio-medical*' model, which deny environmental causality. We acknowledge that there is plasticity between 'nature' and 'the social', and that 'genes take their cue from nurture'. With a critique of the '*Bio-medical*' approach, we have paved the way, having considered the transitions in social theory, the evidence for genetic causality from evolutionary psychology and the social/ethical implications of selected examples of human biotechnology and the Human Genome Project, for an application of the thesis framework to the study of biotechnology in Chapter Five.

Bio-Medical Discourse

Powell and Owen (2005: 27) make the point that *anti-reductionist* social theory 'is a relatively 'new' but methodically eclectic body of theory which analyses the complexity of the tripartite of theory, power and practice'. They draw, as we do here, from the work of Roger Sibeon (1996, 1999 and 2004), and acknowledge the potential of the sensitising framework for theoretical construction/reconstruction, and its emphasis upon methodological generalisations as opposed to substantive generalisations. Reference is made to the work of Owen (2006b), who has modified the framework (employed here, and in particular in Chapter Five for the study of human biotechnology) to include a focus upon *the biological variable*, or evidence for, at least in part, genetic causality. The authors offer the *anti-reductionist* approach as an alternative to *reductionist 'Bio-medical'* approaches identified by Powell and Biggs (2000).

Powell and Owen (ibid) discuss how anti-reductionist theory arises out of a critique of four illegitimate forms of reasoning: *reductionism, essentialism, reification* and *functional teleology*; is anti-relativistic; yet draws heavily from the work of Foucault; and has an ontologically-flexible sensitivity towards a 'post-postmodern' sociology, or what Mouzelis (1991) has termed a 'return to' sociological theory and methodology. This 'new' stance challenges the 'postmodern' theory of fragmentation in society (Seidman, 1994) and rigid explication of human behaviour. Such an approach 'shows that social life is processual which is marked by fluidity, complexity and contradiction' (Powell and Owen, ibid). Indeed, both *anti-reductionist sociology* (Sibeon, 1999, 2004, 2007) and Post-Foucauldian theories (Owen, 2006a, 2007b; Rose and Miller, 1990; Powell and Biggs, 2001) can avoid the *reductionism* of totalising analyses and single unitary explanations. Indeed, both encourage 'contingent or open-ended explanation of practices' (Powell and Owen, ibid). Like Foucault, Sibeon (2004) would reject, 'the reductionist, methodological collectivist and macro to micro theories of bio-medicine' because, 'it views power as outcome of their causal powers as top down: in any power the gains of one side dominates the other' (Powell and Owen, ibid). Following Mouzelis (1991), Powell and Owen (ibid: 28) suggest that a. 'sociological' perspective as opposed to a rigid, *reductionist* 'bio-medical' approach is required in analyses of the relationship between *macro* and *micro* tiers and links, and between temporal impartialities or contingencies between *agency* and *structure*. As emphasised previously, the favoured approach here is one which favours *mutuality* between genes and environment, and an *undersocialised* approach is regarded as equally illegitimate as any *oversocialised* perspective (for example, Gagnon and Simon's (1973) work pertaining to learnt 'sexual scripts') which rejects biological variables out of hand in analyses of causality.

As Powell and Owen (ibid: 31) suggest, 'The bio-medical model represents the contested terrain of decisions reflecting normative claims', and refers to medical techniques and 'social actions' (Hindess, 1988) that privilege a biological and psychological understanding of the human condition', relying upon, ''scientific assumptions' that position attitudes' (ibid). As Arthur Frank (1990: 6) notes, the bio-medical model occupies a privileged position in contemporary culture/society:

Bio-Medicine (occupies) a paramount place among those institutions and practices by which the body is conceptualised, represented and responded to. At present our capacity to experience the body directly, or theorise it indirectly, is inextricably medicalised not socialised.

As Powell and Owen (ibid) go on to suggest, some doctrines of the biomedical model may 'closely reflect the basic sciences while others refer to the primary concern of medicine, namely diseases located in the human body'. Most important is that these beliefs 'hold together' thereby reinforcing one another and forming a coherent orientation towards the mind and body. Indeed, the mind-body dualism had become 'the location of regimen and control' in positivistic, methodological searches for objective 'truth'. Having acknowledged this, it must be emphasised (as has been the case in Chapters Two and Three) that it is sometimes necessary to maintain *dualism* in analysis of phenomena, especially with regard to *agency* and *structure,* and *biology* and *'the social'* in order to avoid the '*conflation*' which Margaret Archer (1988; 1995; 1998) cautions against. Nevertheless, an end product of the misuse of mind-body dualism in the West has been the 'bio-medical model'. In this sense, bio-medicine is based upon the insights of the biological and psychological sciences. By developing an all-encompassing range of bio-medical and totalised discourses, many forms of social injustice can be legitimised as 'natural', inevitable and necessary for the successful equilibrium of the social whole, such as mandatory retirement and the allocation of pensions (Phillipson, 1998).

Powell and Owen (ibid: 29) examine the history of biomedical reductionism, arguing that the model is *reductionist* in attempting to 'perceive and problematise' the subject (in this case the focus is upon human ageing, or social gerontology) as a pathological 'problem' tied to discourses (in this case, of 'decline' and 'dependency'). Thus, as Powell and Biggs (2000) have claimed, problem orientation' is historically configured in 'bio-medical sciences'. This medical science 'problem' approach (again the focus is upon ageing, but the approach can be understood to be applied in a 'broadbrush' sense to phenomena in general) can be related to 'how human subjectivity was structured under occidental modernity' (Powell and Owen, ibid), where, beginning in the 17th and 18th centuries, the 'social sciences', industrial capitalism and bureaucratic politics synchronised novel ways of objectifying 'individuals' and 'populations' in Western societies (Biggs and Powell, 2001). As Powell and Owen (ibid) suggest, the emergence of Western rationality was accompanied by a growth of scientific traditions such as positivism, each 'intellectualising the nature and extent of individuality'. Empirical observation, reason and science are major themes of the modernist project. This is a set of beliefs about validating knowledge that has consolidated modernism. The beliefs emerged from 'the intellectual prognosis of society' during and after the 'Enlightenment' period (Powell and Owen, ibid). The second half of the 18th century, deemed the 'Age of Reason', saw the idea of linear, evolutionary 'progress' elaborated by Kant, Turgot, Condorcet and others that gave rise to modernity (May, 1996). The French Revolution in 1789 – revolution based upon 'reason' – both expressed and gave momentum to this *new consciousness*, and the 'Industrial Revolution' provided its material substance (Powell and Biggs, 2000). This modern world, this new social order, was characterised by a new dynamism, a rejection of earlier traditions, a belief in progress, and the potential of human reason to promote freedom. Increasing rationality would 'enhance social understanding, order and control, justice, moral progress and human happiness' (Powell and Owen, ibid: 30.

Lynn's previously covered concept of 'yuk reactions', objections held towards the use of human biotechnology, especially the examples of human cloning and germline gene therapy (see 'Gene Therapy' subheading in this Chapter's section on 'Further Ethical Concerns and Human Biotechnology') owes something to the *reductionism* of the Cartesian epistemology outlined by Powell and Owen (ibid) above. Lynn dismisses 'emotive' reactions towards certain techniques of biotechnology when applied to human beings as 'feelings of emotional aversion unsupported by arguments based on rational principles' (ibid: 273). In other words, the author echoes the Cartesian belief in the supremacy of 'logical reason' and the illegitimacy of argument based upon emotive variables. As we have hopefully made clear, the approach favoured here is a modification of Sibeon's *anti-reductionism* which rejects modernist *essentialist* and *reductionist* constructs in favour of multifactoral explication and a metatheoretical, 'flexible-but-realist social ontology' (Sibeon, 2004: 17).

Powell and Owen (ibid) refer to 'the notion of transcendence'. Indeed, this idea was used to act as a basis for objective knowledge, 'reinforcing the Cartesian 'method' of existence and cognition and ratifying the need of disembodied experience, yet, dialectically exposing a synthesis of mind and body', where the latter became, 'the obedient rather than the 'prison' of the former' (ibid). As such, and in 'the episteme of transcendence', experience is deemed to be 'real' only if 'deeply entrenched within consciousness and entirely detached from the corporeal' (ibid). Personal subjectivity is thus regarded as a 'threat to the credibility and validity of experiential knowledge' (ibid), and can be transcended only if thawed into the crucible of 'unity', in other words, into the realm of Kant's unified 'transcendental subjectivism' (Husserl, 1990: XX). However, alongside this, differentials, in terms of who may have access to and be able to deploy 'reason' may also 'be seen to have served as a sophisticated legitimisation function between 'scientific experts' and 'subjects' of knowledge' (ibid: 31). For example, given 'the ontological reciprocalisation between repression and freedom, individuals will tend to define themselves via their position or identity within a power relationship such as those of doctor/patient, judge/judged' (ibid).

Powell and Owen's focus here, to reiterate, is upon human ageing, and it posits an *anti-reductionist* alternative to Bio-medical *reductionism*, applies several of the concepts found in Sibeon's original framework and applied here in modified form, and refers to the use of biotechnology to modify and enhance the performance of identity. To recap, the approach here is as much against the *under-socialised, reductionist* approaches of *'the Bio-medical model'* as it is against the rejection of biological causality out of hand as is found in the work of Gagnon and Simon (1973). The approach favoured in the 'new' framework is one of *mutuality* between genes and the social.

Powell and Owen (ibid: 36) consider bio-medicine to be 'normative in its pronouncements embedded as it is' in what Katz (1996) has called ''power/knowledge' relations which prevail in the wider society'. By this, following from Foucault's (1980) relational analysis of power (favoured by Sibeon, 2004; Owen, 2006a, 2006b, 2007a, 2007b] and part of the 'new' framework here applied in greater detail in the following Chapter Five], Katz (1996) does not advocate pitching the blame for the negative position of social actors (in this case, the elderly) in a top-down direction that would exemplify 'downward reductionism' (Mouzelis, 1991). Rather, as Powell and Owen (ibid) suggest, that we are subjected to various multiple forms of power to which we also subject others as an anti-reductionist analysis provides a rethinking of power as 'the outcome of a set of social practices and techniques', that are, as Turner (1989) suggests, given shape by prevailing truth claims about the 'social body' and the conduct of everyday life. Thus, social actors (in this case, the elderly) are subjected to different forms of power which are exercised through techniques of regulation such as governmentality (Rose and Miller, 1990). As Powell and Owen (ibid) suggest, the forms power takes are therefore exemplifying such variability over time and space.

The authors point to 'three ontologically reflexive practices that challenge the hegemony of bio-medical epistemology' (Powell and Owen, ibid). Firstly, and it will be remembered that the context here is one of social gerontology, there is the growing interest in the maintenance of existing good health rather than *reductionist* definitions of 'decline' (Biggs, 1999). This includes the growing market for healthy eating and exercise from midlife onwards. Secondly, there is an increase in the use of notably narrative therapies in later life transcending 'bio-medical', psychological discourses which suggest that the mind 'decays' (Biggs, ibid). Thirdly, Powell and Owen (ibid) refer to, 'the use of biotechnology to modify, and in some cases, enhance bodily performance rather than 'wear out' as an ontologically illegitimate narrative of social action (c f Sibeon, 2004; Shilling, 1993, Featherstone and Wernick, 1995).

As Powell and Owen (ibid: 37) suggest, 'the most notable challenge to a top-down biomedical *reification* of ageing' can be seen in the use and alternative promotion of counselling for older people (c f Biggs, 1993). The focus here (lifecourse counselling and narrative counselling) is upon 're-inventing' oneself in line with current life-priorities. It therefore involves a characteristically different relationship to the personal past than that adopted by traditional *reductionist* bio-medicine. As Powell and Owen (ibid) point out, whereas 'the past had been seen as a repository of experiences that pre-determined choices in the present', the *agency*-focused, ontologically reflexive narrative approach sees personal memory as experiences that can be 'recombined to tell whatever story best fits contemporary identity needs'. Wahidin (2003) makes a similar case for the role of agency in relation to identity (in her case, the 'management of bodies' in a prison context). McAdams (1993) links the popularity of narrative techniques with wider social trends towards a blurred and bifurcated lifecourse. 'Because' states the author, 'our world can no longer tell us who we are and how we should live, we must figure it out on our own' (ibid: 35). As Powell and Owen (ibid) suggest, this time the focus is on the psyche: 'a healthy mind is associated with the capacity to re-author oneself and keep that narrative going – moving beyond narrow and limited discourses of decline of mind and body'.

Against the under-socialised 'Bio-medical model', Powell and Owen refer to the use of new technology to either modify the appearance or performance of identity. Indeed, Donna Haraway's (1991) original reference to cyborgic fusions of biological and machine entities has been taken up by many older people (Longino and Powell, 2004). As Powell and Owen (ibid) suggest, 'the list of bio-technologies available includes virtual identities created by and reflected in the growing number of 'silver surfers' using the Internet as a free-floating form of identity management'. This 'identity management' has been defined by Powell and Biggs (2001) as 'consumer agelessness'. Featherstone and Wernick (1995: 3), against biomedical *reductionism*, claim that it is now possible to 're-code the body itself' as information technologies make available the capacity to alter not just the meaning, but the very material infrastructure of the body. Bodies can be reshaped, re-made, fused with machines, and empowered through technological devices and extensions.

Powell and Owen's critique of 'the Biomedical model' in their words, shows how, 'an anti-reductionist paradigm has transcended explanations of bio-medical theorising' (ibid: 38). According to the authors, 'the bio-medical model has probed, tested and hypothesised' about its subject (here, we will recall, the focus is on social gerontology), constructing 'quantifiable profiles of the bio-psychological and narrowly defined factors deemed to be lying at the root' of the elderly body and mind (ibid). For Powell and Owen (ibid), 'such individualised epistemologies generate problematised discourses into older people's lives' which serve to reinforce the view that it was 'them' rather than 'the problem based reductionist bio-medical narratives that are at fault in 'clumsy-theorising''. As the authors suggests, 'Bio-medicine retains a pivotal position within the modern society' (ibid). Bio-medical narratives produce what Powell and Biggs (2000) have called 'metaphors of pathology', which not only serve to set parameters for debates about their subject (again, here the focus is upon social gerontology) but seep into the public domain, playing a crucial role in discussions about social problems and responses to them. A 're-problematising of the 'aging body' or the re-embedded nature of growing/becoming old should have a place in an anti-reductionist analysis of old age', suggest Powell and Owen (ibid), because 'aging bodies are not as fixed as bio-medical models suggest and are ultimately open ended to contingence and flexibility'.

This Chapter has examined selected examples of human biotechnology and the Human Genome Project in terms of sociological and ethical (particularly eugenic) implications. We have also examined selected perspectives from the 'sociology of science' especially with regard to the Genome Project, and how it 'exemplifies a number of ongoing patterns of change in contemporary science' (Bucchi, 2004: 134), pertaining to the emergence of a 'new' 'Big Science' in the form of collective endeavours carried forth by huge, complex organisations (ibid); the emergence of a 'post academic' science characterised by the intersection of disciplines and by the heterogeneity of the actors involved and by social accountability (ibid); and the emergence of a 'new' 'academic-industrial-governmental' complex which replaces the old 'military-industrial' complex (ibid). Though the intention in this book is *not* to explore issues pertaining to *risk* or *governance* in any great detail, it was necessary to examine briefly the literature on both topics because of their relevance to and association with general sociological interest in human biotechnology. After examining general issues and eugenic issues pertaining to selected examples of human biotechnology we examined the 'Bio-medical model' (Powell and Owen, 2005) in order to emphasise the need to avoid analytical accounts which rely upon biological *reductionism* as well as the *oversocialised* or harshly 'environmentalist' conceptions of the person to be found in the work of Gagnon and Simon (1973). We applied some of the terms, such as *reification, functional teleology* and *reductionism* to a critique of the 'Bio-medical model', but not in the depth that the terms are applied to the study of human biotechnology in order to demonstrate the framework's explanatory potential in the following Chapter.

Chapter Two dealt with the transitions in sociological theory, and made the case that we need to move beyond the relativism and paralysis of postmodern theory. It was argued that a modification of Sibeon's (1999, 2004) *anti-reductionist* framework [itself anti-relativistic, metatheoretical and part of the post-postmodern 'return to' sociological theory/method associated with those such as Layder (1997, 2004) and Mouzelis (1991, 1993, 1996)], to include a focus upon the *biological variable* or evidence from evolutionary psychology/behavioural genetics for a, at least in part genetic basis for some human behaviour alongside *agency-structure, micro-macro and time-space* was a possible 'way forward'. This would entail 'building bridges' between the social and life sciences. Chapter Three codifies the 'new'framework, and provides evidence from evolutionary psychology/behavioural genetics for what Ridley (1999, 2003) has called *'nature via nurture'* – the *plasticity* and *mutuality* of genes and environment in respect to causality. Now, after Chapter Four's examination of issues and implications surrounding human biotechnology, we are ready to apply the metatheoretical framework in greater depth to the study of human biotechnology. All Chapters are interlinked and represent logical stages in the process of developing a framework and finally demonstrating its explanatory potential. Chapter Five, which follows, applies all the terms in the framework to the study of human biotechnology, and this involves covering the literature of various schools including actor-network theories and posthuman approaches which have attempted to address biosocial issues, as well as, for example, identifying examples of illegitimate theoretical reasoning (*reification, reductionism,* etc) in the social science literature pertaining to human biotechnology.

Chapter Five

An Application of the Metatheoretical Framework to the Study of Human Biotechnology

The metatheoretical framework employed in this book has been used to suggest a 'way forward' beyond post-postmodern relativism, in tandem with its application to the study of human biotechnology in the work of Owen (2006b). It has also been used, as a 'Post-Foucauldian' sensitising device in the work of Owen (2006a), and is alluded to in Powell and Owen's (2005) critique of the 'Bio-medical Model' of aging. Here, it is applied in much greater depth to the study of human biotechnology. We apply the framework's terms to the social science literature pertaining to the study of the subject, and also to the work of those such as writers of the 'embodied' school, actor-network theorists and posthuman agency theorists, who have attempted to deal with related biosocial issues.

The Meta-Theoretical Framework

The framework arises out of a critique of the following illegitimate forms of theoretical reasoning:

1. *Reductionism.* This term is included in Sibeon's (1999, 2004) original anti-reductionist framework. Reductionist theories are ones which attempt to reduce 'the complexities of social life to a single, unifying principle of explanation or analytical prime mover (Hindess, 1986a, 1988) such as 'the interests of capitalism', 'patriarchy', 'rational choice', 'the risk society', 'trust', 'the information society', 'globalisation', or whatever' (Sibeon, 2004: 2).
2. *Essentialism.* This term is included in Sibeon's original framework. *Essentialism* is, 'a form of theorising that in aprioristic fashion presupposes a unity or homogeneity of social phenomena' (Sibeon, ibid: 4). This can include social institutions, or taxonomic collectivities such as 'white men', 'the middle class', etc.
3. *Reification.* This term too is included in Sibeon's original framework. *Reification* is the 'illicit attribution of agency to entities that are not actors or agents' (ibid). In Sibeon's (ibid) view, and echoed in the 'new', modified framework, an *actor* is an

entity that, 'in principle has the means of formulating, taking and acting upon decisions'. Sibeon's original, non-reified definition draws upon Harre's (1981) concept of agency and Hindess's (1988: 45) 'minimal concept of actor', which specify that for an entity to be regarded as an actor, it must be capable of reaching and acting upon decisions. As Sibeon (ibid) has made very clear, on the basis of such a non-reified definition, there are two types of actors, namely individual human actors and Hindess's concept of 'social actors' (1986a: 115). 'Social actors' include organisations such as government departments like the Home Office, organised pressure groups etc, and committees such as the Cabinet and micro groups such as individual households (Sibeon, ibid: 5). These 'social actors' have been termed 'supra-individuals' by Harre (1981: 141).

4. *Functional Teleology.* Again, this term forms part of Sibeon's original framework. It may be defined as, 'an invalid form of analysis involving attempts to explain the causes of social phenomena in terms of their effects, where 'effects' refers to outcomes or consequences viewed as performances of 'functions' (Sibeon, ibid: 6). Sibeon's definition draws upon the work of Betts (1986: 51), and the argument is that, if there is no evidence of intentional planning by actors 'somewhere, sometime' (Sibeon, ibid), then it is a teleological fallacy to engage in explication of the causes of phenomena *in terms of their effects*. As Sibeon (ibid: 6-7) suggests....

> All too often, reductionist theorists – including some who subscribe to varieties of 'critical social theory' – begin with a current social or cultural item (a social practice of some kind or, say, a law, a welfare or health system, or a public policy), then attempt to 'work backwards' and claim, without any demonstration of intentional planning by previous actors, that the item came into being 'because', in the view of the theorist, it accorded with the interests of say, a taxonomic collectivity such as 'the upper class' or 'white people' or 'men'.

5. *The Oversocialised Gaze.* This term forms part of the 'new' modified framework alongside the previous four illicit forms of theoretical reasoning. This 'new' term refers to harshly 'environmentalist' accounts which are characterised by a strong antipathy towards genetic, or partially genetic explication. The term has been applied to the work of Gagnon and Simon (1973) by Owen (2006a, 2006b, 2007a), in order to criticise the symbolic interactionists' theory that there is no 'natural', sexual drive in human biological make-up. As we have seen in Chapter Three (the section entitled, Incorporating the Biological Variable), contrary evidence can be found in the work of Hamer and Copeland (1999: 163), who have clearly shown how genes influence our sexual desire, how often we have sexual congress and 'help make us receptive to the social interactions and signs of mutual attraction that we feel instinctively and now call love'.
6. *Genetic Fatalism.* This is another 'new' term, now incorporated within the framework employed here. As Owen (2006a, 2006b) shows, the term refers to a widespread tendency within social science to *equate genetic predisposition with inevitability.* As we have seen from the work of Ridley (1999, 2003), in Chapter Three (the section entitled, *Building Bridges*), it is a mistake to view the genes involved in human behaviour as immutable. As Ridley (1999: 153) cogently

suggests, 'genes need to be switched on, and external events – or free-willed behaviour – can switch on genes'.

In addition to these 'cardinal sins', the framework focuses upon the *metatheoretical* formulations of *agency-structure, time-space* and *micro-macro*, in addition to notions of Foucauldian *power,* the 'new' term, *the biological variable,* and favours a *dualism* rather than a *duality of structure*. These terms are briefly re-explained as follows. It will be recalled that a *metatheory* is, as Sibeon (2004: 13) suggests....

intended to inform and hopefully improve the construction of substantive theories and the design of empirical studies. Metatheorists are largely concerned with ontological questions, including the following. What, in general terms, is 'society'? What sorts of things exist in the social world? If there are such things as actors or agents, what sort of things are they? Are activities and society indivisible ('two sides of the same coin'?) and so tightly melded together that (as claimed in doctrines of ontological duality) it is impossible to separate them? Does it make sense to employ a stratified social ontology that refers to micro and macro spheres or 'levels' of society, or is micro-macro – as Foucault, Elias and Giddens claim – a spurious and misleading distinction?

Such questions are the concern of *metatheory*, and of *metatheoretical* concepts such as *agency-structure*. In short, metatheories and meta-concepts are designed to, 'equip us with a general sense of the kinds of things that exist in the social world, and with ways of thinking about the question of how we might 'know' that world' (Sibeon, ibid: 13).

7. *Agency-structure*. These are important meta-concepts which refer to significant aspects of social reality. In Sibeon's original framework, and as applied here in the 'new' modified framework, the conception of *agency* is a non-reified one, in which actors or agents are defined as entities that are, in principle, capable of formulating and acting upon decisions. *Structure* refers to the 'social conditions' (Hindess, 1986a: 120-1), or the circumstances in which actors operate, including 'the resources that actors may draw upon' (Sibeon, ibid: 54). *Structure* then, may refer to discourses, institutions, social practices and individual/social actors.
8. *Micro-macro*. Another meta-concept which 'refers to differences in the units of and scale of analyses concerned with the investigation of varying extensions of time-space' (Sibeon, ibid). *Micro* and *macro* should be viewed as distinct and autonomous levels of social process. Apparently, Sibeon was influenced by Layder's (1994, 1997) argument to the ends that events in social life, at one level, do not determine events at anther level, although 'there may be contingently produced and contingently sustained empirical connections between levels.' (Sibeon, ibid: 55).
9. *Time-space*. This meta-concept refers to, 'significant but neglected dimensions of the social' (Sibeon, ibid: 166). As Sibeon has made clear, the term reflects concerns with temporality and spatiality. Classical social theorists (Durkheim, for example) have tended to regard time, as 'social time', distinct from 'a natural essence' (Sibeon, ibid). However, 'the question of how differing time-frames – including those associated with the macro-social order and those with the micro-social – interweave is a complex matter' that relates to debates pertaining to *dualism* versus *duality* (Sibeon, ibid). Arguably as the author also suggests, there is a connection too with

Derek Layder's (1994, 1997) (meta) theory of *social domains*, in which the links between objective and subjective domains involve a mingling of 'two different sets of time-frames, namely those associated with (macro) institutional time and those with (micro) situated activity' (Sibeon, ibid).The spatial dimension of society here receives attention in relation to notions of indeterminacy and to spatial variation. Sibeon (ibid: 156) refers to 'the neglect of spatiality' in sociology as opposed to studies of temporality. Soja (1989), too, has made this same point. However, as Sibeon (ibid) makes clear, there has been a certain amount of recent interest in the significance of locale, contingency and spatial variation, notably in community studies (Bulmer, 1985, 1986).

10. *Power*. Like Layder (2004), Sibeon acknowledges the *multiple* nature of power. For Sibeon (ibid: 145), his (meta) theoretical precepts lead to a view that, 'power exists in more than one form; in particular, there are objective structural (including systemic) forms of power, and agentic power', a term used to refer to 'the partly systemic and partly relational and potentially variable capacity of agents to shape events in a preferred direction'. This modified notion of Foucauldian *power* favoured by Sibeon has also been applied by Owen (2006a) in the course of his development of a *Post-Foucauldian sociology of aging*, and by Powell and Owen (2005), in their critique of the *Biomedical Model* as employed in social gerontology.
11. *Dualism*. Sibeon's original framework favours *dualism* rather than notions of duality of structure. In doing so, he is also in the company of Layder (1994, 1997, 2004). As Sibeon has suggested (1997a: 72), like Giddensian *structuration* theory, Foucault's work has a tendency to 'compact agency and structure together instead of treating them as a dualism'. Owen (2006a: 186) has employed *dualism* in favour of *duality*, and makes the point that, 'This Foucauldian tendency (1972, 1980a, 1982) to compact *agency* and *structure* together, to collapse distinctions between the two, results in what Archer (1995) calls 'central conflation'. Owen (ibid: 186-187] goes on to argue for the use of dualism in his own work along the following lines....

> The idea of *dualism* as opposed to duality of structure is favoured in the Post-Foucauldian framework, because it is felt that in social analysis, *agency-structure* and *micro-macro* should be employed as dualisms that refer to distinct, relatively autonomous phenomena. For example Shilling's (1993) analysis of the human body as simultaneously 'biological' and 'social' serves to collapse the terms into an amalgamated 'whole' within which elements cannot be separated.

12. *The Biological Variable.* This 'new' term reflects Owen's (2006a, 2006b, 2007a, 2007b) respectful criticism of Sibeon's original *anti-reductionism* along the lines that it neglects biological or partly biological causality in explanations of social 'reality'. As we have seen in Chapter Three, entitled *A Post-Postmodern Metatheoretical Framework?,* we should regard the *biological variable* as the evidence from evolutionary psychology and behavioural genetics for a, at least in part, *biological* basis for some human behaviour. We should keep Ridley's (1999, 2003) notion of '*nature via nurture'* firmly in mind when focusing upon *biological variables* in analysis. This refers to the 'feedback loop' which embraces both genes and environment, acknowledging plasticity and mutuality. Genes, as we have seen from

the work of Ridley (ibid) and other evolutionary psychologists in Chapter Three, predetermine the broad structure of the brain of *Homo Sapiens*, but they also absorb formative experiences and react to social cues. In short, nurture depends upon genes, and genes require nurture. It will be recalled that in Chapter Three, we examined the evidence for a biological basis for certain human behaviour. In particular, the evidence that language is innate and specific to our species was examined with reference to the work of Hamer and Copeland (1999), Ridley (2004), Dunbar (1996) and others and the work of Hamer and Copeland (ibid), and Bogaert and Fisher (1995) in which evidence that sexual drive is genetically-determined as opposed to being a cultural and historical construction (Gagnon and Simon, 1973; Foucault, 1980b) was also examined. Combined with additional evidence from the work of those such as Harris (1998) which challenges *nurture determinism* with hard evidence from studies in the field of behavioural genetics upon personality traits, the decision was made to include *the biological variable* – evidence for genetic or partly genetic causality as a concept in the 'new' modified metatheoretical framework.

Now, the task is to demonstrate the framework's explanatory potential by identifying examples of the illegitimate forms of reasoning listed previously within the literature and approaches to studying human biotechnology within contemporary social science and to examples of related bio-social issues such as the 'bridge building' between social and biological sciences. This will also involve an application of the metatheoretical concepts to the literature and approaches. Selected examples, including the Human Genome Project are examined, and prominent 'schools' of social science which attempt to 'build bridges' between the social and life sciences/conceptualise issues pertaining to the use of human biotechnology, such as *Actor-Network theory* and *Posthuman* approaches are looked at through the lens of the 'new', metatheoretical framework. As Delanty (2002: 1) has cogently suggested, the social sciences 'cannot afford to ignore the challenge that is coming from the new genetics', and 'this will involve rethinking some of the epistemological assumptions of modernist social science'. Delanty (ibid) has usefully pointed to, 'a possible opening within the sciences for a reorientation for sociology and in a way that the new genetics can be addressed in a serious way'. This lies in 'a qualified version of constructivism' (ibid). The argument put forward in this book, supported by the work of Owen (2006a, 2006b, 2007a, 2007b) and Powell and Owen (2005) is that another 'opening' may lie in a 'bridge building' exercise in which we acknowledge the *mutuality* and *plasticity* of genes and environment. This, of course, involves acknowledging *the biological variable,* or the evidence from evolutionary psychology/behavioural genetics demonstrated in Chapter Three, for biological causality. Yet, predisposition, as Ridley (1999, 2003) cogently suggests, need not be viewed through the lens of *genetic fatalism* – genes respond to social cues and are not immutable. Arguably, a post-postmodern, metatheoretical framework which employs Ridley's (ibid) *'nature via nurture'* approach is capable of contributing both to the 'return to' sociological theory and method associated with Layder (1994, 1997, 2007), Sibeon (1999, 2004, 2007) and others, and also to the study of human biotechnology. The 'pay off' will hopefully lie in the framework's explanatory potential in terms of further theoretical development.

REIFICATION

Reification, the 'cardinal sin' of illegitimately attributing *agency* to entities which are not actors can be found within the literature of social science pertaining to the study of human biotechnology and its sociological implications. As we have seen in Chapter Two, the work of Owen (2006b:911) has suggested that the Human Genome Project is an entity that is *not* an actor, and that to argue otherwise would be to engage in *reification.* As Owen (ibid) makes clear....

> The government of a group of 'nation states' may decide that it is desirable to create a project such as the Human Genome Project that is 'empowered to take decisions and act on the collective behalf of its member states' (Sibeon 1996: 57). In this case, the Human Genome Project may be regarded as a *supra-national organisation.*

Owen (ibid) suggests, we need to keep firmly in mind Hindess's (1986: 115) formulation of social actors (a crucial component of the framework's methodology inherited from Sibeon's original anti-reductionism), in which Hindess defines an actor as a 'locus of decision and action where the action is in some sense a consequence of the actor's decisions'. There are those, however, such as Wilkie (1994: 187) who, in a discussion of the Genome Project and its implications and the impact of new genetic knowledge upon individuals/social institutions engages in reification when he claims that institutions 'have a history and their own inertia and 'will to live''. Additionally, the author reifies 'society' too, when he claims that, 'society needs to demarcate those things which will be permitted and to list those which, for the moment, ought to be forbidden' (Wilkie, ibid: 188).

Cazillis (2001) has identified the problem of embryo reification in the use of stem cells. The author claims that we are facing ethical problems related to the moral status of the embryo, genetic engineering and ownership rights on living organisms raised by modern human biotechnology and its use. As stem cells are, 'the most recent consequence of the over-accelerating biological process, it is not surprising that they are triggering most of the heated debates in the life sciences section' (Cazillis, ibid: 1). Such debates have to steer a path through two main dangers, according to the author in a EUROPA, European Commission Research Centre article. On one hand, there is the possibility of taking hasty decisions (under pressure from international competition), and on the other hand, there lies the possibility of missing the opportunity 'for therapeutic progress able to relieve suffering and save lives' (ibid). The author argues that embryonic stem cells pose acute problems. They can only be obtained from embryos aged approximately one week, at the blastula stage. The embryo at the blastula stage is approximately nine days old. At the centre of this cell mass there lies a cavity filled with liquid, and the few cells it contains – embryonic stem cells – are the 'cause of all the excitement among biologists' (ibid). Research on embryos is extremely controversial, as we have seen in Chapter Four, irrespective of the stage of the embryo's development. As Cazillis (ibid) shows, a report by British stem cell experts, in this case from the UK Department of Health in 1990, stressed that a significant body of opinion believes that it is morally and ethically unacceptable to use an embryo for any form of research whatsoever. This is for the reason that some feel the embryo should be recognised as having a full human status immediately it is conceived. At the other end of the spectrum, the Department of Health researchers found that others claim that the embryo neither requires nor merits any kind of 'special status'. Others, the report shows, accept the 'special status' of the embryo as a 'potential' human being, but maintain that the respect due to the embryo increases as it develops and grows, and that this 'respect' has to be weighed against any potential benefits of research.

As Cazillis (ibid) points out, some countries do finance research upon human embryos, others refuse to engage in it, and some ban the practice outright. In the United Kingdom, where public financing accepts research along existing lines, but not the creation of new ones, the private sector is subject to the rules laid down by each Member State (ibid). As the author suggests, in cases where such research is permitted, there is the question of where the embryos actually come from. 'Researchers who have developed existing stem cell lines (the exact number is not known, but it undoubtedly runs into dozens) have used surplus embryos' (ibid). These were created then conserved by the process of freezing with the intention to use them in in vitro fertilisation, but then they ceased to be part of a parental project either because the couple had parted company or the IVF was successful. They were thus destined to be destroyed. As Cazillis (ibid: 2) suggests, 'It is not known how many embryos of this kind are available, and if a major research activity were to develop, there are some who inevitably envisage creating in vitro embryos specially for scientific purposes' (ibid). As Cazillis makes clear, this would pose great ethical implications and would be seen as 'one more step in the direction of embryo reification'. The author does not appear to view embryos (which some view as 'potential' human beings) as social actors, hence the use of the phrase reification. There is neither the scope nor the intention in this book to debate ethical issues in great depth, merely to highlight those that obviously emerge during any study of human biotechnology. However, Cazillis (ibid) certainly poses an interesting question. Strictly speaking, if an embryo is unable to formulate and act upon a decision, under the definition employed here in the 'new' modified framework, it is clearly reification to regard the entity as an actor. However, it is for theologists and those who specialise in medical ethics to determine the answer as to whether an embryo should be accorded with the 'special status' referred to previously.

Another controversial area of human biotechnology is the transfer of somatic nuclei, 'sometimes known as 'therapeutic cloning'' (Cazillis, ibid). This involves transferring the nucleus of a 'normal' cell to a previously enucleated egg, with the aim of 'creating' an embryo carrying a patient's genes in order to extract stem cells which are compatible with his/her immune system. In this case too, Cazillis (ibid) points out that embryo reification may apply. What is more, the door may be opened to, 'reproductive cloning or, in other words, the universally condemned practice of creating an embryo carrying the genes of a single individual and then bringing it to its full development' (Cazillis, ibid). Present research is taking place within a very varied, changing legal framework. As the author suggests, questions are being raised about the 'ownership' of cell lines and the tissues they are able to 'create' and eventually transplant (the cell banks will soon be a social reality) and 'on the patentability of the products and techniques which originate during such research' (ibid: 3).

Newman and Brody (1988) writing upon psychosocial and ethical concerns surrounding the 'new' human reproductive biotechnologies, argue that infertility has always represented a form of denial of family continuity, which is undoubtedly a topic of psychological consequence for many individual people and families. As the authors suggest (ibid: 1), 'traditional ways of incorporating a new generation for those who were barren have included sharing of children, adoption of 'foundlings', and fictive kinship'. All of these strategies have resulted in 'anonymous birth parenthood' and thus the reification of social parenthood.

However, at the present point in time, the availability/acceptability of the technologies of contraception/abortion to avoid unwanted pregnancy, mean that the number of children available for adoption has diminished considerably in the so-called developed world. At the same time, new techniques of human biotechnology for treating infertility have been developed that combine 'donor genetics, multiple gestators, and sophisticated medical interventions so that infertile couples can have their 'own' infant, thereby achieving symbolic and social continuity' (ibid).

Newman and Brody (ibid) refer also to an area of concern pertaining to the moral/ethical domain which is part of 'the cultural context of doctor-patient relationships'. It refers to, 'what both doctors and patients regard as good, proper, right, correct or desirable; it applies equally to the converse – what they regard as bad, improper, wrong, incorrect or undesirable' (ibid). Such perceptions are arguably determined by the overall cultural, including religious and legal, context of medical practice (ibid). As the authors go on to suggest, 'in most of the Western world they favour, at least as ideals, the principles of autonomy (or self-determination)', and 'justice (as in equality of access to resources and competing claims to scarce resources), and beneficence in medical care (the value that the patient, the person in need, not the family, the doctor or the state) comes first' (ibid). Newman and Brody (ibid: 2) are of the opinion that, 'taken together they express the Kantian view of respect for persons', in this case, the patient. However, such ideals are inevitably compromised. Thus, certain religious groups view the principle of patient autonomy with considerable alarm. As the authors suggest, 'in the past this was expressed mainly by limiting access to contraceptive technology on grounds that coitus was intended only to produce children, and that such technology would foster promiscuity' (ibid). Today, the pro-life movement express similar alarms, whilst working towards the prohibition of abortion and towards granting foetuses the status of persons, thus interfering with the clinical judgement of physicians, the experience of patients and research to improve both, in the view of Newman and Brody. Here, we can apply the metaconcept of reification again, as the reference to the 'pro life' desire to award special status to foetuses may be interpreted as embryo reification as was identified in the work of Cazillis (2001). It seems difficult to regard a foetus as an actor in the sense we employ in the framework, that is to say, an entity that is, in principle, capable of formulating and acting upon decisions. However, as stated previously (in the treatment of Cazillis's work in this Chapter), we must leave the finer points of any ethical discussion of the status of embryos to medical ethics experts or theologists.

At this point, it is necessary to consider the important matter of non-agency. We need to keep this firmly in mind as in what follows, we examine the 'schools' of Actor Network theory and Posthumanism which both have conceptions of the actor which are greatly at odds with the non-reified conception favoured both by Sibeon (2004) and in the 'new' modified framework. To reiterate, we are applying Sibeon's original conception of an actor as an entity that, in principle, has the means of formulating and acting upon decisions to the study of human biotechnology. 'Schools' such as Actor Network theory and Posthumanism both, arguably, attempt to redefine the relationship between the social and nature and are regarded as related to sociologies of science. It is argued here that their reified conceptions of the actor make the work of the schools less useful as conceptual frameworks with which to study human biotechnology or to 'build bridges' between the social and life sciences. In relation to the 'cardinal sin' of reification, Sibeon (ibid: 122) argues that, 'agency is not synonymous with social effects'. There are lots of phenomena possessing social effects (the author provides the example of a conditioning effect upon agency), but they are not necessarily actors. Sibeon (ibid) makes the point that many different forms of phenomena, such as 'magazines, rain, the HIV virus, the River Nile, motor cars, the Internet, and the Moon' have 'social consequences', but to label them actors equates to the illegitimate attribution of agency. Sibeon (ibid) identifies reification in the work of Law (1991a) pertaining to his conception of agency. Law's work is seen as closely influenced by Actor Network theory here, alongside structuralism, poststructuralism etc. According to Law (1991b: 173-4), 'an agent is a structured set of relations with a series of (power) effects'. As Sibeon (ibid) points out, this does not include strategies and intentions, and he asks how, ''social relations' formulate, take and act upon decisions'? Sibeon makes it clear that he cannot agree with either Callon's use of poststructuralism and actor-network theory in regard to Callon's (1986: 204) claim that scallops are actors, and Callon's (1991: 142) claim that the Chernobyl nuclear plant is an actor. As Sibeon (ibid) suggests, such reified claims can be found in some of the more mainstream areas of social science literature. For example, he cites the work of Clegg (1989: 200) who claims that computer systems display agency, and that 'agency may be vested in non-human entities as diverse as machines, germs... and natural disasters'. As Sibeon (ibid) says, such views are misplaced and should be rejected.

Recently, there have been attempts to extend Actor-Network theory beyond studies of technology, power and organisation of the body into dialogue with feminisms, anthropology and psychology by Law et al (1999). Williams-Jones and Graham (2003) have claimed that social, ethical and policy analysis of the issues arising from gene patenting and commercial genetic testing using human biotechnology, is enhanced by the application of Actor-Network theory. They appear to be very much influenced by the approaches of Callon and Latour (1981), Callon (1986) and Latour (1987). As Bucchi (2004) has said of this 'school', for the proponents, science has two faces, 'like the Janus of Roman mythology: on the one hand there is 'ready made' science; on the other, science 'in the making' or research', and while it is 'the task of epistemology to analyse the characteristics of the former', it is 'the task of the sociology of science to study the latter (ibid: 70). Williams-Jones and Graham in their recent (2003) application of Actor-Network theory to the ethical analysis of commercial gene-testing, appear to be particularly influenced by the distinction (perhaps seen to best effect in Latour (1987)) between human and non-human actors. As Bucchi (ibid: 72) suggests, for the proponents of Actor-Network theory....

> A research colleague, a bibliographical citation in a paper, an apparatus which yields a microscope image, a company willing to invest in a research project, a virus that behaves in a certain way, the potential users of a technological innovation: all these are allies in the process that transforms a set of experimental results and statements or a technological prototype into a 'black box': a scientific fact or a technological product.

Bucchi (ibid: 70) also suggests that *Actor-Network theory* can be viewed, 'as an attempt to expand the explanatory capacity of the microsociological approaches to science'. Williams-Jones and Graham (ibid) take the theories and go on to suggest that there is potential in transferring *Actor-Network theory's* 'flexible' nature (ibid: abstract) to an applied heuristic methodology for 'gathering empirical information and for analysing the complex networks involved in the development of genetic techniques'. The authors explore three concepts in their *New Genetics and Society* paper – *Actor Network Theory*, translation and drift – and apply these to the case of Myriad Genetics and their commercial BRACA analysis genetic susceptibility test for hereditary breast cancer. If we apply the meta-concept of *reification* to their work, we can clearly observe that the authors commit the 'cardinal sin' of reification – the illegitimate attribution of *agency* to entities which are not actors. Williams-Jones and Graham (ibid) treat the susceptibility test as an active participant in socio-technical networks. This is rather like Callon's (1986) treatment of scallops. The authors suggest (ibid) that the test, 'interacts with, shapes and is shaped by people, other technologies, and institutions', and that 'such an understanding enables more sophisticated and nuanced technology assessment, academic analysis, as well as public debate about the social, ethical and policy implications of the commercialisation of new genetic technologies'. It is difficult to see how the authors could possibly justify their claim, because it is surely based upon a *reified* notion of *agency* on the part of 'the test'. How a test for the susceptibility of breast cancer can be capable of formulating and acting upon decisions is not explained by Williams-Jones and Graham (ibid).

Recently, Pickering's [1995a, 1995b) *Posthumanism* has suggested some original conceptual resources for theorists. As Jones (1996: abstract) suggests, Pickering 'tenders nothing less than a fundamental reordering of social thought', by offering a 'decentred variant of constructivist sociology of science', which invokes the concept of 'material agency', and seeks to 'redefine the relationship between "Nature" and "Society", while dismissing 'the "humanist bias" inherent in sociological inquiry'. As Sibeon (2004: 122) has made clear, he repudiates Pickering's (2001) notion of *posthuman agency*, 'which attributes agency to machines and physical objects', as *reification*. Pickering, as Jones (ibid: 291) points out allies himself 'in qualified fashion' with the *Actor-Network* programme in science studies championed by those such as Callon (1986) and Latour (1988).

As Jones (ibid) argues, *Actor-Network theory* 'finds its conceptual footing in the semiotics of French structuralist and poststructuralist thought'. It is, in essence, a form of 'antihumanism', depicting sociological explication/description as 'illegitimate impositions of modernist categories' (ibid). Callon and Latour (1981) contend that moderns having been subject to scientific representations, must abandon human-centred forms of inquiry/expression if they are to develop what Jones (ibid) terms, 'genuinely holistic appreciations of social life in its material settings'. Callon and Latour (1981) suggest instead a method of theorising associations between people and things that eschews all established ways of speaking about such relations. A decentred 'technoscience' is given primacy, in which 'society is no longer distinguished from the scientific, the technological, or the natural' (Jones, ibid). The object of analysis becomes a 'unified field of "heterogeneous" networks that stitch together elements drawn from each of these "counterfeit" categories' (ibid). By 'disrupting commonsense appreciations of "mankind", "nature" and "society" – indeed by rejecting such classifications altogether', Callon and Latour (ibid) seek a 'new' approach to investigating the ways in which social life is 'enmeshed in technological projects' (Jones: ibid). Jones makes the point that Pickering 'concurs, generally, on the need for this strategy' and notes that Pickering (1986: 561) has suggested that 'sociology has traditionally focussed upon human individuals and groups as the locus of understanding and explanation', and to remedy the problem as he sees it, now recommends a, '*posthumanist* displacement of our interpretive frameworks'.

Jones (ibid: 292) makes the point that it is the concept of 'nonhuman agency', the idea that natural objects 'act' as human beings are able to do, that is 'the linchpin in the antihumanist argument'. For Jones, Pickering's point of view appears to be that neither sociology nor poststructuralism, nor *Actor Network theory* 'can relate or distinguish with complete adequacy the "performative" (ie concrete and temporal) qualities of things material, human and nonhuman' (Pickering, 1996b: 9-15). Jones (ibid) poses the question, 'has he got this right?' For Jones, the answer is in the negative. As far as we are concerned, in relation to the meta-concepts of the framework, the answer lies in the negative too. We employ here, a tightly-drawn, *non-reified* concept of the actor that borrows from Sibeon (2004). Sibeon (ibid), it should be noted, acknowledges social actors (organisational actors) in addition to individual actors, who, in the latter case, must be able to (in principle) formulate and act upon decisions. Pickering (1995b) is arguing for a *material agency* – material relations between people and things can be considered in terms of *human agency*. Sibeon (ibid: 147) usefully discusses the issue of whether machines are actors, in his case, specifically computer systems. The author makes the point that 'inside many aircraft there are 'automatic pilots' and in all major airports air traffic control systems are computerised'. He goes on to mention that in the field of medicine, 'software is available for assisting in the diagnosis of illness', and that the 'industrial application of robotics, in car manufacturing, for example, is highly advanced'. However, as Sibeon importantly makes clear, 'in each of these cases' the computerised decision-making is wholly 'dependent on hardware that is made and maintained by humans'. The computer software is 'bounded' by parameters that are determined and monitored by human programmers' (ibid). Sibeon does not deny the possibility that 'advances in computing hardware, software and bio-engineering might in future prompt some revision of the anti-reified conception of actor (agent)' (ibid). However, at this stage, it is merely *reification* to concur with Pickering's conception of *material* or *non-human agency*.

Jones (ibid: 293) emphasises that *Posthumanism* 'refers to deliberate rejections of "scientific realism" (the belief that nature alone determines the ways in which scientific representations are developed and refined), *and* "social realism" (the idea that questions of scientific fact are never answered definitively by nature but are decided instead by human interactions)'. As the author suggests, in this sense, it resembles poststructuralist attempts to 'decompose and negate the epistemic warrants of all natural and social scientific inquiries' (ibid). Poststructuralism, as Jones suggests, does not provide any resources for 'conceptualising material phenomena as they take on form and substance in "real time"'. In their rejection of both social and natural sciences, they provide no substitutes. Here, as Jones (ibid) suggests, *Actor-Network theory* differs from poststructuralism. 'They *do* propose a materialist alternative to scientific (and commonsense) descriptions', claims Jones. For the author (ibid: 294)...

> Latour suggests that modern science, interpreting the world as it does in Cartesian and Kantian terms, paints an inadequate portrait of human and nonhuman *realities* (1993). Through the opposition of subject and object, society and nature, science misrepresents the world. Its categories are carved from reality in objectionable ways: science fails to capture the essential continuity or "symmetry", of people and things. As a remedy, actor-network theory recommends that analysts grant to material objects the semiotic equivalent of human agency. But the kinds of semiotics that Latour proposes differs from the poststructuralist variety in important ways (see Latour 1993: 63-64). Actor-Network theory asserts that signs must be

interpreted not only in relational terms – that is, within the logic of a closed semiotic system – but now also in terms of empirical conditions.

Pickering appears to borrow from the 'decentred perspective' of the *Actor-Network* scheme, but as Jones (ibid: 295) points out, he 'cannot take seriously notions of metaphysical or nonhuman intentionality'. On one hand, as Jones (ibid) shows, Pickering (1993: 565) has claimed that, 'we humans differ from nonhumans precisely in that our actions have intentions behind them'. Yet, on the other, he still attempts to 'preserve a decentred outlook' (Jones, ibid) when he recommends that we, 'acknowledge a role for nonhuman – or material, as I will say – agency in science' (1993: 562). Pickering, as Jones (ibid) cogently suggests, 'is searching for a way to displace the human agent without envisioning a semiotic subject that vanishes entirely from the field of free action'. In other words, under the 'new' framework's definition of actor, Pickering is still engaging in illegitimate theoretical reasoning – *reification.*

Jones (ibid: 297) makes what many might consider a cogent point when he suggests that, 'it is a mistake to spend time puzzling over the meaning of nonhuman agency in posthumanism or actor-network theory – the important question is whether individuals are granted some measure of autonomy, or alternatively, are represented as subjects of structural logics'. This point is important because 'the preservation of asymmetric human agency' is the real purpose of posthumanism, and the reason Pickering distinguishes his approach from that of Callon and Latour's (1981) *Actor-Network theory.* As Jones (ibid: 297–298) suggests, 'the conceptual viability of material resistance depends on the retention of purposeful human activity'. In *posthumanism,* resistance tends to 'emerge' only *as* resistances *to* autonomous human agents (the tendency here is still firmly towards *reification*). As Jones (ibid: 298) argues, resistances cannot be reconciled with antihumanism, and 'this is the structure/agency problem all over again', arising from the *Posthumanist* tendency to draw 'conceptual resources from discordant theoretical tradition'. For Jones, Pickering's project might be better served by, 'reconsidering the enduring questions in the human search for self-knowledge', because 'posthumanism is a form of forgetfulness' (ibid: 306). Jones (ibid) explains himself thus; 'by surveying established traditions of social thought, unproductive detours through "new improved" brands of inquiry can be avoided'. Jones refers to recent attempts to suggest that the foundations and purposes of sociological theory and method are 'suspect' by poststructuralists and postmodernists, and entreats us to 'accept no substitutes' for the discipline and methods of sociological tradition. One certainly cannot accept the *reified* conception of the actor employed by *Posthumanists* and *Actor-Network theorists*, and the framework employed here is intended to be a contribution towards the post-postmodern 'return to' sociological theory and method. However, that does not mean a return to 'sociological business as usual'. It must be emphasised here that a metatheoretical, flexible ontology is favoured, and an approach which combines 'sociological realism' with an acknowledgement that genes *do* influence behaviour. This is not the same as a recommendation that we return to the *oversocialised* conceptions favoured in some strands of 'mainstream' sociology. An example here would be Giddens's (1993) apparent denial of human instincts in the sense of unlearned patterns of behaviour.

In what follows, we apply the meta-concept/ 'cardinal sin' of *essentialism* to the study of human biotechnology and related literature which attempts to redefine the relationship between 'the social' and 'nature'. Our task here, to recap, is to demonstrate the conceptual usefulness of the 'new' modified framework.

ESSENTIALISM

Essentialism as employed in Sibeon's (2004: 4) system refers to, 'a form of theorising that in aprioristic fashion presupposes a unity or homogeneity of social phenomena, such as the law or some other social institution' or taxonomic collectivities such as 'women' or 'white people' etc. The term essentialism is used here to refer to the illegitimate attribution of homogeneity to social phenomena on a priori grounds. This is distinct from 'the uses of this term which are to do with the question of whether phenomena - social categories such as 'women' for example – have real essences or are socially – constructed' (ibid: 197).

Regarding the latter interpretation of essentialism, Nelkin (1994) has noted the 'spiritual' aspects of 'genetic essentialism' in which genes are imbued with all the various properties of the person or self in certain literature pertaining to human biotechnology and specifically the Human Genome Project. Nelkin, an observer of the 'popular' understanding of genetics, draws attention to promotional metaphors characterising the HGP as 'the Holy Grail'. Alexander (2003) develops this line of analysis further, looking at how biotechnology has become 'a new religion', and exploring the prophets, disciples and 'spiritual creed' of biotechnology. Fleising (2001) looks at the 'genetic essentialism' inherent in the assumption that different meanings for DNA have ethical/social consequences. The author suggests that 'genetic essentialism' as an interpretive meaning for DNA encapsulates values of materialism and autonomy that make it compatible with capital accumulation. However, despite the interesting nature of the interpretation of essentialism to be found in the work of the above authors, we are concerned here with the concept of essentialism as clearly outlined by Sibeon (2004) and detailed above.

With this firmly in mind, we can proceed to 'apply' the term essentialism to the study of human biotechnology and the literature pertaining to it. One recent example of essentialism in the sense that we employ it here, lies in the work of Buctuanon (2001), in her paper pertaining to the 'globalisation of biotechnology', in the journal New Genetics and Society. Sibeon (2004) refers to Albrow's (1996: 91-2, 94–5) criticisms of 'essentialist notions of 'globalisation', in the sense that essentialism 'does not regard the degree of homogeneity or heterogeneity of social phenomena as an empirical variable for investigation, but rather presupposes on theoretical or political grounds a necessary 'unitariness' of phenomena (Sibeon, ibid: 4).

Here, Buctuanon (ibid: 25) appears to commit the 'cardinal sin' of essentialism by regarding biotechnology as homogenous phenomena. Her paper examines, 'a contradictory development in the era of globalisation wherein country-specific economic and socio-political institutional environment limits the global flow of technological knowledge and information, particularly in the biotechnology sector'. However, here 'biotechnology' does not appear to be referred to in any specific sense, and we are none the wiser as to which form of biotechnology or which 'techniques' or 'technologies' the author refers to. Buctuanon (ibid: 26) claims to be examining, 'the present globalisation process of biotechnologies', and suggests that, 'biotechnologies tend to agglomerate in developed economies, leaving the pure markets and limiting state involvement'. However, we have no idea as to whether this refers to gene-enhancement products, reproductive technology or whatever.

Very often essentialist theory is linked to or is a corollary of a reductionist theory (indeed, it tends to be reductionism that underpins or nurtures the other illicit forms of reasoning). With this point in mind, we can also see how Buctuanon (ibid: 26), in addition to assuming a priori essentialist unitariness of biotechnology as phenomena, also posits a reified notion of the state. The author's claim (ibid) that, 'the state has assumed an active responsibility of promoting the well-being of its firms by organising the innovation system in which the latter have been embedded' is arguably a case of the illegitimate attribution of agency to an entity which, like taxonomic collectivities, should not be regarded as an actor.

Sibeon (1996: 34) makes the point that although essentialism is clearly related to reductionism it has a tendency to be more specifically linked, 'to the reductionist notion that taxonomic collectivities – such as "women" – are a relatively homogeneous category comprised of individuals (individual women, in this instance) who have more or less common (and "objective") interests' that are 'given' to them because of their structural position in the taxonomic collectivity. The author suggests that essentialism can be found in abundance in contemporary feminisms, for example (ibid). As he goes on to show, essentialist feminist theories of which Elliott and Mandell's (1995) work is cited as an example, mistakenly suppose that, ''women" as represented in the theory is a social category that is empirically "real"', in other words a category given by biology/society rather than the 'product' of social construction or a feminist theoretical construct (ibid). With these points in mind, it is therefore possible to identify the 'cardinal sins' of essentialism within certain feminist critiques of human biotechnology. Hanmer (1993) refers to how the Western 'birth control movement' was influenced by Galtonian eugenic ideas and their political expression, in addition to how the movement may be perceived as a grass-roots movement to enable women to space their children in order to protect their health. In doing so, she arguably engages in essentialism and reification when she refers to 'this double edge of control by women and of women by the state remains an unresolved issue today' (ibid: 231). Hanmer refers to women as if they are a homogeneous category whilst attributing agency to 'the state', which is an entity most definitely not an actor according to the definition employed by Sibeon, and in the 'new' framework outlined here. She also draws attention to a critique of human biotechnology common to feminisms, in which the notion of choice for women is challenged by 'arguing that the pressures on women to have children, but only when it is socially acceptable to do so' (ibid: 232) are so great that it is pointless to discuss issues of 'choice'. Again, the emphasis is upon women as an essentialist monolithic block. 'Women are seen as having no option about whether or not to have babies', and 'motherhood is socially compulsory, even in countries with coercive population control policies' are phrases (ibid) which further indicate Hanmer's essentialist tendencies. Hanmer identifies another 'line of attack' which criticises human reproductive technologies from feminist standpoints, for being, 'about something very different from what is claimed', in other words, 'not about helping women to have children, but rather being an aspect of genetic engineering closely associated with new developments in biotechnology' (Kollek, 1990, Shiva, 1988). This includes the attempt to map the entire human gene pool, beginning with the Human Genome Project, to 'improve' human embryos through genetic testing and the addition/subtraction of genetic material (Kaufmann, 1998; and Leuzinger and Rambert 1998). Again, the emphasis in Hanmer's work is upon the implications for 'women' as a category which is, in an essentialist way, empirically "real" and homogeneous. Hanmer goes on to refer to feminist literature such as that of Bridenthal, Grossman and Kaplan (1984); Degener (1990); Ewing (1988); Kaupen-Haus (1988); Schleiermacher (1990); and Zimmerman (1990), which suggests that 'modifying and eliminating people before birth gives a new twist to eugenics' (ibid: 234). However, her claim that 'it can also be seen to be about men as a class taking control of women's reproductive activities' is surely a case of essentialism. Here, Hanmer appears to refer to both 'women' and 'men' in essentialist terms as homogenous social phenomena.

In what follows, we apply the meta-concept of Functional Teleology to the study of human biotechnology and literature pertaining to the field. This is, as was the case with reification and essentialism, to demonstrate the framework's explanatory potential and the usefulness of specific meta-concepts in relation to the study of biotechnology.

Functional Teleology

In Sibeon's (2004: 6) original scheme, and as employed here in the context of the 'new' meta-theoretical framework, *Functional teleology* is, 'an invalid form of analysis involving attempts to explain the causes of social phenomena in terms of their effects'. This is where 'effects' refer to *outcomes* and *consequences* 'viewed as performances of 'functions'' (ibid). Sibeon appears to have been influenced by Betts (1986: 51), whose work has suggested that it is illegitimate to try to explain the causes of phenomena 'in terms of their effects', in the absence 'of a demonstration of intentional and successful planning by actors' (Sibeon, ibid). Sibeon (ibid) cautions against the *reductionist* tendency (particularly prevalent amongst those who 'subscribe to varieties of 'critical social theory'') to attempt to 'work backwards'. That is to say, to claim without evidence of intentional planning by previous agents, that a certain phenomena came 'into being' because it accorded with, 'the interests of, say, a taxonomic collectivity' such as 'the middle class' or 'white men' etc. As the author cogently suggests, *teleology* is flawed 'by a problem of logic' in the sense that the causes that 'create' a social/cultural phenomenon must *predate* the existence of the phenomenon (ibid). Consequences or effects can surely only occur in respect of phenomena already in existence.

With this concept of *functional teleology* in mind, we can 'apply' it to the study of, and literature pertaining to, human biotechnology. As Sibeon (1996: 20) has suggested, 'Marxism and radical feminism, have been bedevilled by the problem of teleological explanation'. Powell and Owen (2005: 28), drawing upon Sibeon, (in a critique of the 'Bio-medical' model of human aging) refer to 'Marxist gerontology' which 'attempts to explain 'old age' as an effect of or in terms of needs of capitalist social systems'. Hanmer (1993: 232), drawing upon feminist critiques of human biotechnologies, suggests that the 'new reproductive technologies' are used to uphold traditional, patriarchal 'notions of motherhood and femininity via the selection of those who should or should not be mothers'. Here, we have an example of *functional teleology*, in which it is suggested, in the absence of evidence of conscious planning by social actors, that 'patriarchal forces' are acting to maintain the inequality endured by women in society. Hanmer goes on to argue (ibid) that, 'the social factors of class, 'race' and ethnicity, and male domination determine differential treatment of women'. The author claims, in another example of illegitimate *functional teleology*, that in Britain, we see 'those social forces at play with the new reproductive technologies' and that 'women who are not dependent on men are largely deemed to be unacceptable for motherhood by the medical gatekeepers and the state'. Aside from the *functional teleology* inherent in an explanation which offers no evidence of conscious planning by social actors, and refers to 'social forces', there we have also a *reified* notion of 'the state'. Hanmer refers to the fact that human biotechnologies (in this case, reproductive ones) are 'expensive' (ibid) and access is often limited to those who can afford to pay. That may be true, but to argue as Hanmer (ibid) does, along the lines that 'patriarchal rule' is somehow behind the divisions between women that may arise because of unequal access to human biotechnologies based upon fiscal wealth, is surely to employ *functional teleological* explanations again. Hanmer (ibid: 234) refers to another 'line of attack' criticising the new reproductive technologies 'for being about something very different from what is claimed'. In other words, IVF for example, should not be seen as a technology which helps women to have children, 'but rather being an aspect of genetic engineering' closely associated with advances in human biotechnology (ibid). This, Hanmer argues, includes the Human Genome Project, the attempt to map the entire human gene pool, and attempts to 'improve' human embryos through genetic testing. Hanmer (ibid) again employs illegitimate *functional teleological* explication when she argues that, 'it can also be seen to be about men as a class taking control of women's reproductive activities as husbands/male; partners/fathers; scientists/medical practitioners; businessmen; and governmental leaders'. Again, there is no evidence provided of any conscious planning by agents nor of how 'taking control' of women's reproductive activities accords with the interests of taxonomic collectivities such as 'husbands', businessmen' or 'scientists' etc. As Sibeon (2004: 7) has suggested, such accounts are 'flawed by a problem of logic in so far as the factors ('causes') that bring a social or cultural item into being must necessarily predate the existence of that item' whereas effects or consequences can, of course, only occur 'in respect of an item that is already in existence'. The author goes on to suggest that *functional teleology*, 'conflates causal and functionalist explanation' in the sense of seeking to explain causality in terms of the item's 'functions' (ibid).

Next, we look at the 'cardinal sin' of *reductionism*. It should be emphasised at this point that all the 'sins', or illegitimate forms of theoretical reasoning identified here are not confined to any particular theories or paradigms. Rather, as Sibeon (ibid: 5) suggests, they, 'crop up almost everywhere across the theoretical landscape'. It should be noted that Sibeon refers to *reification, reductionism, essentialism* and *functional teleology* here. Sometimes, as we have previously encountered in the work of Hanmer (1993: 232), an author may engage in more than one form of illegitimate reasoning. In this particular case, Hanmer is employing an explanation which relies upon *functional teleology* in the sense of suggesting that 'social forces' (without providing any evidence of conscious planning by social actors) manipulate the new human reproductive biotechnologies, and relying upon *reification* in the sense of suggesting that 'the state' formulates and acts upon decisions pertaining to which women are 'deemed to be unacceptable for motherhood'.

Reductionism

In Sibeon's original framework, and in the modified framework employed here, we regard a *reductionist* theory as one which, 'illegitimately attempts to reduce the complexities of social life to a single, unifying principle of explanation or analytical prime mover' (Hindess, 1986a, 1988)' (Sibeon, 2004: 2). This might be 'globalisation', 'the risk society', 'patriarchy' or 'capitalism' etc. It must be emphasised here that *reductionism* differs from *essentialism* in that the latter 'sin' is a form of theorising which presupposes a unity of social phenomena. *Reductionism* also differs from *genetic fatalism*, which is a 'new' term included in the framework employed here, *not* employed by Sibeon, and which refers to a tendency to equate genetic predisposition with *inevitability* (in other words, accounts which ignore the evidence, supplied by those such as Ridley (1999, 2003), that genes can be 'switched on' by environmental stimuli). As Sibeon suggests, a key feature of *reductionist* general theories is an 'ontological inflexibility' in which they rest upon 'a priori assumptions' (ibid), and offer unitary, substantive explication. However, we need to be cautious in our approach to the term 'reductionism'. As the author suggests, to reject 'a priori theoretical commitment to analytical prime movers' is not to suggest that there are, 'never situations where a very small number of factors (or perhaps only one) may have causal primacy' (ibid: 3). However, we should regard this as an empirical question rather than as something (in advance) which is predetermined on the basis of reductionist, unitary explication. Additionally, Sibeon stresses that we should note the difference between, 'non-reductive multi-factorial explanation' and what he has termed 'compounded reductionism' (ibid). The latter form of theorising involves 'attempts to combine or synthesise two or more reductionisms' such as 'capitalism' and 'patriarchy' (ibid). Thirdly, we are entreated to avoid what Sibeon terms 'deferred reductionism', in which reductionist explication is 'postponed or deferred' (rather than removed from analysis) (ibid). The author gives the work of Farganis's (1994: 15-16) feminist critique of postmodern theory, in which Farganis 'acknowledges that social class, age, ethnicity, and sexuality are important dimensions' of social life/existence, but argues 'in the last instance' it is always the case that gender is, 'the ultimate or primary axis of social life' as a cogent example of 'deferred reductionism'.

Buctuanon (2001: 26) appears to engage in *reductionism* (in the sense of reducing phenomena to a single, unifying principle of explication) in her article concerning the 'globalisation of biotechnology' in the journal, *New Genetics and Society*. Here, the author examines, 'the present globalisation process of biotechnologies', and 'the trends in the dispersion of knowledge and information in biotechnology through the activities of large transnational corporations' (ibid). Buctuanon appears to rely here upon a 'classic' *reductionist* model of globalisation, in which it is suggested there is a very 'high level of global connectedness of phenomena' (Sibeon, 1996: 153). However, as Sibeon (ibid) has suggested, a problem with such approaches lies in their tendency to, 'exaggerate the scale and/or the intensity of globalisation' (ibid). Such models of globalisation processes 'underplay the highly uneven impact of globalising tendencies' and they also imply that such processes are 'irreversible' (ibid). As Sibeon (ibid) suggests, 'this orientation tends towards reductionism, and also towards essentialism insofar as it erroneously assumes the existence of a social process (globalisation) that is relatively unified'. Bretherton (1996: 12) may be nearer the mark when he suggests that globalisation is, 'a net of overlapping processes that are neither inexorable nor irreversible, the impact of which varies in intensity and is highly differentiated in effect'. Arguably, Buctuanon's work upon the globalisation of human biotechnologies also entails reliance upon a *reified* notion of the state. The author refers to how the global flow of technologies 'tends to agglomerate in developed economics particularly in the US', where a mixture of 'economic choices, social regulation and state action fostered institutional environments that facilitated the development and commercialisation of biotechnologies' (ibid: 25). Also emphasised is the need for 'a re-thinking of the role of the state' in technological and economic development, as the author considers 'the state' a 'political entity that structures the innovation system in order to promote the well-being of its firms' (ibid). To re-emphasise, it is the view here that we should regard as *reification* any accounts which suggest that 'the state' possesses the agency to formulate and act upon decisions.

O'Brien (1981, 1989) in her feminist analysis of the politics of human reproduction engages in *reductionist* theorising by arguing that the 'commodification' of life processes is structurally related to the interests of capitalism and the societies in which it has developed. For O'Brien, it is logical in any capitalist economic system that human reproduction and life processes should be 'commodified', as ownership is the highest social priority of individuals. These social relations of science and medical technology 'were being developed in previous socialist societies in relation to women's bodies and life processes generally' (Hanmer, 1993: 241). This is, suggests O'Brien (ibid) because capitalist systems can restructure ideology in order to serve their economies and power structures. Arguably, O'Brien's theory also incorporates a *reified* notion of the 'capitalist system'. Hanmer (ibid: 240) interprets O'Brien (ibid) as suggesting that 'gender domination varies in its forms as societies alter over time' and now it is, 'becoming possible to intensify control over women through highly-developed technologies based on reproductive biology'.

Authors such as Turner (2004), Nelkin (1994), Hall (2003), Alexander (2003) and West (2003) in various ways arguably engage in *reductionist* theorising along the lines of conceptualising human biotechnology 'as religion'. Each author appears to reduce the complexities of human biotechnology to an ultimate, single, unifying principle of explanation. As stated, in these cases, the analytical 'prime mover' (as Hindess, 1986a, 1988) would call it, is religion. Nelkin (1994) has noted how, since the early days of the Human Genome Project, genetics has served not just as a scientific research programme, but also as a theology or, as Turner (2004: 659) puts it, 'a quasi-religious belief system'. Nelkin (ibid) has also noted the 'spiritual' aspects of 'genetic essentialism' in which genes are imbued with properties of the person. It is important to make clear the distinction between this concept of *essentialism* and the one formulated by Sibeon and employed in the framework here. In contrast to Nelkin's (ibid) interpretation, *essentialism* is a meta-concept of the 'new', modified framework, and must be viewed as, 'a form of theorising that in a prioristic fashion presupposes a unity or homogeneity of social phenomena' (Sibeon, 2004: 4). Alexander (2003) develops Nelkin's analytical theme, exploring the 'prophets', 'disciples' and 'spiritual creed' of human biotechnology. Hall (2003) suggests that within the belief systems of posthumanists and other technological enthusiasts, biotechnology offers the possibility of life extension, even, as Turner (ibid) notes, 'more fancifully', the idea of immortality within this world. The author suggests that for 'technological enthusiasts' biotechnology is a surrogate religious system in which somatic cell nuclear transfer, stem cell research, gene therapy regenerative medicine etc will enable scientific researchers to slow or reverse aging processes and permit human beings to live prolonged lives. West (2003) suggests that 'Technoevangelists' have referred to the prospect of achieving immortality through downloading human minds into computers or utilising cybernetic technologies to develop human-machine cyborgs with replaceable parts. Turner (ibid: 659-660) has recently argued that whilst 'traditional religious cosmologies place eternity on some other plane of existence', the 'biotechnological fantasies locate immortality here on earth'. For the author, 'Biotechnology provides a surrogate religious narrative' which 'instead of imagining a day of rapture when the dead rise from their graves to begin eternal life', the enthusiasts 'imagine an era when medical technologies provide a renewable, largely imperishable body' (ibid).

In what follows, we apply the meta-concept of *Genetic Fatalism* to the study of human biotechnology and literature pertaining to the subject, in order to demonstrate the explanatory potential of this 'new' term. Sibeon's original anti-reductionist framework does not include this concept, which as Owen (2006a, 2006b, 2007a, 2007b), Owen and Powell [2006] and Powell and Owen (2005) define as the tendency within social science to associate genetic predisposition/determinism with *inevitability*. As we have observed in Chapter Three, Ridley (1999: 53) has shown how there is a *mutuality* and *plasticity* between genes and environment, and how it is a mistake to deem genes as 'immutable' because they, 'need to be switched on, and external events – or free-willed behaviour – can switch on genes'.

GENETIC FATALISM

As previously stated, this meta-concept refers to the tendency to equate genetic predisposition/determinism with a fatalistic *inevitability*. We have examined evidence from the work of Ridley (1999, 2003) against this 'cardinal sin' of illegitimate theoretical reasoning, which suggests a great degree of *plasticity* and *mutuality* between genes and environment. Our focus here concerns not only literature pertaining to the study of human biotechnology, but as stated at the beginning of the Chapter, work which attempts to deal with bio-social issues. For example, in the previous section on *Reification*, we examined the work of Actor-Network theorists. Here, it is relevant to recall the 'vigorous counter-attack against the idea of genetic influences on behaviour' led by Richard Lewontin and Stephen Jay Gould referred to by Ridley (1999: 306) and mentioned in Chapter Three. As Ridley notes, a favourite slogan of those involved in the counter-attack against the ideas of E.O. Wilson was, 'Not in our genes!' Perhaps at the time it was 'a plausible hypothesis to assert that genetic influences on behaviour were slight or non-existent' (Ridley, ibid). However, the work of Rose, Kamin and Lewontin (1984) may be 'no longer tenable' in the light of studies in behavioural genetics which offer cogent evidence that, 'genes do influence behaviour' (Ridley, ibid). It will be recalled that in Chapter Three we have examined the compelling evidence from authors in the fields of Evolutionary Psychology and Behavioural Sciences, such as Hamer and Copeland (1999), Enard et al (2002), Lai (2001), Dunbar (1996), Pinker (1994), Harris (1998), Bogaert and Fisher (1995), Curry (2003) and Cosmides and Tooby (1997) amongst others for an, at least in part, genetic basis for human behaviour.

As Ridley (ibid: 307) argues, there is no 'escape from' determinism by 'appealing to socialisation' because, 'either effects have causes or they do not'. As the author suggests....

> If I am timid because of something that happened to me when I was young, that event is no less deterministic than a gene for timidity. The greater mistake is not to equate determinism with genes, but to mistake determinism for inevitability. Said the three authors of *Not in our genes*, Steven Rose, Leon Kamin and Richard Lewontin, 'To the biological determinists the old credo "you can't change human nature" is the alpha and omega of the human condition'. But this equation – determinism equals fatalism – is so well understood to be a fallacy that it is hard to find the straw men that the three critics indict.

It should be emphasised here that Ridley (1999, 2003) is arguing for the concept of 'Nature via Nurture' here, as opposed to simplistic 'nurture' *or* 'nature' arguments. At no point does Ridley argue for *purely* genetic causality, and in fact acknowledges that, even after the discoveries of evolutionary psychology and behavioural genetics pertaining to the influence of genes upon behaviour, 'environment is still massively important, probably in total more important than genes in nearly all behaviours'.

Schaffner (a Professor of Medical Humanities at George Washington University, USA) writing in 2000 argues for a *'genetic-environmental determinism'*, which appears to be very similar to Ridley's 'Nature via Nurture' model. However, it could be argued that Schaffner (ibid: 326) engages in *genetic fatalism* when he describes his paper as being, 'critical of *genetic* determinism'. Arguably, the author is making the same mistake that Ridley identifies as equating determinism with *inevitability*. In a sense, Schaffner need not posit *'genetic-environmental determinism'* as an alternative to *genetic determinism*, as Ridley has shown how such genetic causality/determinism 'works' in conjunction with environmental influences in any case. Nature depends upon genes and genes need nurture.

Interestingly, Schaffner draws attention to the considerable amount of philosophical and scientific inquiry directed, over the course of the last hundred years, at issues of reduction and reductionism. Here it should be emphasised that Schaffner (ibid: 302) refers to *reductionism* in the context of the Human Genome Project, and closely associates the concept with the notion of 'genetic determinism' rather than in the sense that we employ the term *reductionism* (as a 'cardinal sin') in the 'new' framework in relation to theories which attempt to reduce the complexities of the social world to a single, unitary explanation. By 'genetic determinism', Schaffner means that, *inherited* genes are taken to be the principal drivers of an organism's behaviour (ibid). The author examines the extent to which there are purely, or even primarily, 'genetic' explanations of behaviour. Reference here is made to Kupferman's (1992) review of Kandel, Schwartz and Jessel's *Principles of Neural Science* (later published in 1995). The book which Schaffner (ibid: 304) calls, 'a bible of sorts in neuroscience', begins by making the point that behaviour in all organisms is shaped by the interaction of *both* genes and the environment. The relative importance of the two variables varies, but even 'the most stereotyped behaviour' can be 'modified' by environment and the most plastic behaviour (such as language) is influenced by innate factors (Kupferman ibid: 987). As Schaffner (ibid) suggests, Kupferman goes on to focus on aspects of behaviour that could conceivably be inherited, and on the processes of interaction between genes and environment affecting behaviour. Thus, as Schaffner (ibid) makes clear, the point about the incompleteness of any exclusively genetic basis for the explanation of behaviour is taken as a major premise in the scientific community. The search for *pure, single gene* genetic explanations for behaviour, as Schaffner (ibid: 323) suggests, may be conceptualised as the 'strongest of "strong" genetic programs', but the concept is revealed to be incoherent. As the author suggests, 'any behaviour requires an organism acting in an environment' (ibid). The organism may thus be regarded as the 'product' of many genes which are required to 'build' the nervous system 'integrated within the organism that is productive of behaviour'. However, as Schaffner (ibid) suggests, the Huntington's Disease (HD) paradigm, 'in which we appear to have a single gene that is productive of HD in all current environments' has a powerful hold on both geneticists' thinking and media/popular thought. The reference Schaffner (ibid) makes is to the work of Rubinsztein et al (1996) in the *American Journal of Human Genetics*. However, as Schaffner (ibid: 318) suggests, 'The Huntington's Disease (HD) paradigm of a (prima facie) single dominant 100% penetrant gene' may be more 'the exception rather than paradigmatic' for the search for such single genes for bipolar disorder and schizophrenia, 'has *not* been successful'.

As Schaffner (ibid: 317) shows, it is the case that there have been advances in the molecular genetics of, 'a variety of *single gene* somatic disorders'. The prototype may be sickle-cell anaemia which dates back to the 1949-1957 years, 'when it was characterised by Pauling and Ingraham as a molecular disease due to a single mutation' (ibid). In 1983, the author shows, the gene for Huntington's Disease was localised to chromosome 4, 'though the specific locus of this gene was difficult to determine, and had to wait until 1993' (ibid). Genes for polycystic kidney disease, retinoblastoma and cystic fibrosis 'were mapped in 1985, and this latter gene was localised and cloned in 1989' (ibid). Recently, as Schaffner indicates, 'genes for two types of inherited colon cancer and two forms of breast cancer have been localised and sequenced' (ibid).

The genetics of mental disorders, however, 'has not fared as well', though some important/controversial advances have been made in 'the behavioural realm' (ibid). As Schaffner shows, the finding that there was a gene associated with schizophrenia on chromosome 5, 'turned out to be fallacious'. Kennedy et al (1988) support Schaffner's claim. As Kelsoe et al (1989) has made clear, a report identifying a simple gene for manic-depressive disorder on chromosome 11 had to be withdrawn from the literature. Schaffner (ibid) makes the point that the only 'single-gene-behavioural trail' example of which he is aware involves 'a point mutation in the structural gene for monoamine oxidose A' that was associated with a 'vaguely defined' impulse/disorder/aggression phenotype reported by Brunner et al (1993a; 1993b). Although Schaffner acknowledges the methodologically 'sound' science employed by Brunner (ibid), the latter does not characterise the mutation as an 'aggression gene'. Brunner (ibid: 160) comments that, 'behaviour should and does arise at the highest level of cortical organisation, where individual genes are only distantly reflected in anatomical structure'. As Schaffner (ibid: 318) suggests, recent discoveries in personality genetics include, 'an identification of a gene, DRD4, for "novelty-seeking", but it should be noted that the group reporting a confirmation of that gene noted that this was only a partial – and not a single gene kind of explanation'. The group, according to Benjamin et al (1996: 83) suggested that, 'DRD4 accounts for roughly 10% of the genetic variance, as might be expected if there are ten or so genes for this complex, normally distributed trait'. As we saw in Chapter Three (the section 'Incorporating the biological variable'), Hamer and Copeland (1999: 180) suggest that D4DR may not be regarded as a "promiscuity gene", but it 'determines the way certain brain cells respond to dopamine, which in turn influences a person's reaction to novel stimuli'.

Schaffner (ibid: 326) closes his paper by suggesting that his investigations have not, 'offered any arguments to undercut a combined (and historically-based) *genetic environmental* determinism. This appears to be a similar position to that employed by Ridley (1999, 2003) and favoured in the 'new' modified framework employed here. In Chapter Three (the section entitled 'Incorporating the Biological Variable'), we have seen how Newton (2003: 29), similarly to the work of Rose, Kamin and Lewontin (1984), appears to engage in *genetic fatalism* in his treatment of Walter Cannon's (1914) work. Newton (ibid) also appears to dismiss previous work by Benton (1991: 17) which suggests that Cannon sought to conceptualise life-forms 'holistically' Newton (ibid) instead argues that Cannon's arguments are based upon 'reductionistic' accounts which explain human behaviour, 'chiefly in terms of instincts'. As we saw previously, Cannon's (ibid: 264) suggestion that, 'the emotion of fear is associated with the *instinct* for flight, and the emotion of anger or rage with the *instinct* for fighting or attack' is arguably less of a 'reductionistic' and 'desocialised' account than Newton (ibid) appears to think. It would appear more likely that Cannon is actually arguing for what Ridley (2003: 4) has termed 'Nature via Nurture', a scenario in which genes are reacting to social cues rather than a form of *fatalistic* determinism, in this case the 'cues' being the 'cause' of the emotion of fear.

Wilkie (1994) falls into the trap of equating any explanation which relies upon genes with immutability and *inevitability*. The author refers to how, in his view, the rise of environmental movements since the 1960s has a 'bearing on the attractiveness of biological (which almost without exception means genetic) explanations for human conduct' (ibid 181). Wilkie (ibid) refers to how 'sociobiology' appeals to a kind of romanticism which in turn appeals to nostalgia for 'mythical' ideas pertaining to harmony between 'human biological nature' and cultural form in its attempts to explain modern social life. Wilkie (ibid: 181-182) goes on to argue that, 'some proponents of the Human Genome Project might protest that such a borrowing of genetic explanations of human traits and behaviour is illegitimate', but on the other hand, a few of the project's 'foremost proponents' such as Walter Gilbert (a Nobel Prize-winning geneticist) have remarked that, 'when we know the complete human genome we will know what it is to be human'. Wilkie (ibid: 182) contrasts the difference between 'assumptions underlying sociobiology' and the Human Genome Project (we know, of course, that the human genome was sequenced in 2001). The difference lies in the 'alterable' nature of the composition of the human genome as perceived by 'geneticists' and the sociobiological view that 'the iron hand of the genes' is 'immutable' and 'fixed' (ibid). Wilkie, writing in 1994, appears unaware of the 'reforms from within' sociobiology attempted by the ex-Harvard scholars, John Tooby and Leda Cosmides. The work of the latter authors is (particularly their 1997 work *Evolutionary Psychology: A Primer)* drawn upon in Chapter Three in which we examine the evidence from evolutionary psychology and related fields for an, at least in part, genetic basis for some human behaviour. As Ridley (2003: 245) points out, Tooby and Cosmides, in their Chapter 'The Psychological Foundations of Culture' in Barkow et al (1992) argued that the expressed behaviour of a human being, 'need not be directly related to genes, but the underlying psychological mechanisms could be'. So, as Ridley (ibid) suggests, Tooby and Cosmides could eschew the idea that we can 'search for 'genes for war'', and also 'the contrary dogmatic insistence that war is a pure product of culture written on the blank slate of impressionable minds'. Rather, the authors argue for 'psychological mechanisms in the mind, placed there by natural selection acting in the past upon sets of genes that predispose most people to react to some circumstances in warlike ways' (ibid). This, Tooby and Cosmides (1992) christened 'evolutionary psychology'. As Ridley (ibid) makes clear, it is arguably an attempt to 'fuse the best of Chomsky's nativism – the idea that the mind cannot learn unless it has the rudiments of innate knowledge' with the 'best of sociobiology's selectionism: that the way to understand a part of the mind is to understand what natural selection designed it to do'. For Tooby and Cosmides (ibid), the entire developmental programme thus evolves, and each programme (for an eye, a kidney etc) requires the integration of perhaps thousands of genes but crucially, the presence of 'expected environmental cues' (Ridley, ibid). This, then, may be viewed not as the 'immutable', 'fixed' ideas of 'the iron-hand of the genes'; the *genetic fatalism* fearfully referred to in Wilkie's (1994: 182) account of the implications of the Human Genome Project, but rather 'a subtle mixture of nature and nurture that studiously avoids pitting the two in opposition to each other' (Ridley, ibid). It should be emphasised here that this approach, termed 'nature via nurture' by Ridley, is the one which was argued for in Chapter Three, and the one which it is suggested here is best suited to be combined with the meta-concepts of Sibeon's (2004) original framework. *Genetic fatalism*, as has been demonstrated here, is a 'cardinal sin' which needs to be studiously avoided in theoretical analysis pertaining to issues around human biotechnology and related areas involving possible linkage between life and social sciences. Together with the *oversocialised gaze* and the *biological variable, genetic fatalism*, is a 'new' meta-concept which, as we have seen at the beginning of this Chapter, forms part of the 'new' framework. It is to the concept of the *oversocialised gaze* that we must now turn our attention. Again, we examine the explanatory potential of the meta-concept in relation to the study of human biotechnology/related biosocial issues.

THE OVERSOCIALISED GAZE

The Oversocialised Gaze refers to accounts which are harshly 'environmentalist' and antipathetic towards genetic or partially genetic explanations of human behaviour. The accounts may even reject genetic variables altogether, as Giddens (1993: 57) appears to do when suggesting that, 'human beings have no *instincts* in the sense of complex patterns of unlearned behaviour'. As we have seen in Chapter Three (the section entitled 'Incorporating the Biological Variable'), there is cogent evidence for *instinctive*, unlearned patterns of human behaviour in the work of authors such as Ridley (1999, 2003), Hamer and Copeland (1999), Pinker (1994) and Dunbar (1996) especially regarding the idea that, 'grammar is innate' (Ridley, 1999: 104).

Also examined in Chapter Three (the section entitled, 'Duality Versus Dualism Revisited') was the tendency of writers of the 'embodied school' or 'sociologists of the body' to engage in *oversocialised* approaches. Tim Newton's (2003: 29) criticisms of Sahlins' (1972) and Pollock's (1988) approaches (in the area of 'building bridges' between the social and life sciences) centre around the latter authors' respective portrayals of stress as the product of 'natural' instincts. In Newton's view, 'such arguments present a *crude dualism*' which reduces complex social 'problems' to 'outmoded' conceptions of 'the biological body' (ibid). He then goes on to argue for a 'putative non-reductionistic 'biological sociology''. As noted in the section of Chapter Three referred to above, it is hard to envisage a credible 'biological sociology' in the age of the Human Genome Project which denies human instincts altogether. It is not appropriate here to rehearse all the evidence provided in Chapter Three for the *instinctive* component of human language ability, sexuality etc. However, it could be argued that Newton (ibid) appears to be unaware of the fairly recent evidence provided by Ridley (1999) for an account of stress in human beings which posits the '*nature via nurture*' approach to causality as favoured in the 'new' metatheoretical framework applied here. Newton's arguably *oversocialised* concept of stress emphasises the influence of complex social factors whilst rejecting any reference to 'natural' instinctive, genetically-based behaviour. Ridley (ibid) arguably shows how it is indeed possible to include both genetic *and* environment variables in an explanation of stress in human beings.

As Ridley (ibid: 149) acknowledges, stress in human beings is 'caused' by 'the outside world', and 'short-term stressors cause an immediate increase in epinephrine and norepinephrine', which are the hormones that make the human heart beat faster, the feet go cold etc. These hormones act to prepare the human body for 'fight or flight' in emergencies. Stressors that last for longer activate a different pathway that results in a much slower, but more persistent increase in cortisol. Cortisol is 'used in virtually every system in the body' and is a 'hormone that literally integrates the body and the mind' by altering the configuration of the brain' (ibid). As the author suggests, 'one of cortisol's most surprising effects is that it suppresses the working of the immune system', and those who have shown the symptoms of stress are 'more likely to catch colds and other infections', because one of the effects produced by cortisol is the reduction of white blood cells (ibid: 149-150). Ridley goes on to show how the relationship between genes and environment is one of *mutuality* rather than an *oversocialised* conception which rejects genes and instinct in favour of a solely social explanation of the kind suggested by Newton (2003) in his treatment of the work of Sahlin (1972) and Pollack (1988). Ridley (ibid: 150) explains how the hormone cortisol reduces the 'activity, number and lifetime of lymphocytes', or white blood cells, showing how cortisol does this by switching genes on…

> It only switches on genes in cells that have cortisol receptors in them, which have in turn been switched on by some other triggers. The genes that it switches on will then switch on other genes and so on. The secondary effects of cortisol can involve tens, or maybe even hundreds, of genes. But the cortisol was only made in the first place because a series of genes was switched on in the adrenal cortex to make the enzymes necessary for making cortisol.

As Ridley suggests, 'you cannot produce, regulate and respond to cortisol without hundreds of genes', practically all of which 'work' by 'switching each other on and off'. In white blood cells, 'control is almost certainly involved' in 'switching on' the gene known as *TCF*, on chromosome ten, which enables *TCF* to make its own protein, 'whose job is to suppress the expression of another protein called interleukin 2' which in turn, is a 'chemical that puts white blood cells on alert to be especially vigilant for germs' (ibid). Thus, cortisol suppresses the 'immune alertness of white blood' and renders the person, 'more susceptible to disease' (ibid). The author goes on to pose an interesting question: 'Who's in charge? Who ordered all these switches to be set in the right way in the first place, and who decides when to start to let loose the cortisol?' (ibid). The answer does not lie with the genes solely, even though 'the differentiation of the body into different cell types, each with different genes switched on' is a genetic process 'at root' (ibid). Genes are not the 'cause' of stress, and as Ridley makes clear, 'the death of a loved one, or an impending exam, do not speak directly to the genes' (ibid). They are, in essence, 'information' processed by the human brain. Is the human brain 'in charge'? On one hand, as the author suggests, the hypothalamus of the human brain, 'sends out the signal that tells the pituitary gland to release a hormone that tells the adrenal gland to make and secrete cortisol' (ibid: 151). The hypothalamus takes 'orders' from 'the conscious part of the brain which gets its information from the outside world' (ibid). However, as Ridley suggests, this 'answer' will not suffice 'because the brain is part of the body'. Instead…

> The reason the hypothalamus stimulates the pituitary which stimulates the adrenal cortex is not because the brain decided or learnt that this was a good way to do things. It did not set

up the system in such a way that thinking about an impending exam would make you less resistant to catching a cold (ibid).

In answer to what *did* 'set up the system', Ridley suggests that 'natural selection did', and 'somewhere down the cascade of events that is the production, control and reaction to cortisol' stress-prone people 'must have subtly different genes from phlegmatic folk' (ibid). Who or what is ultimately 'in charge'? Ridley's answer is that 'nobody is in charge', rather an 'intricate, cleverly designed and interconnected system' in which nurture depends upon genes, and genes require nurture in an elegant mutuality. This is, arguably, far removed from the *oversocialised* explanation for stress posited by Newton (2003), or that which the author identifies as '*crude dualism*' in the work of Sahlin (1972) and Pollack (1988).

In Chapter Three (the section entitled, 'Incorporating the Biological Variable'), we saw how Owen (2006a) has identified the work of Gagnon and Simon (1973) and Michael Foucault (1980b) as examples of accounts which rely upon the *oversocialised gaze*. As Owen (2006a: 190) suggests, 'Foucault's (1980b) argument that sexuality is a socio-cultural creation, that sexuality as we know it is the production of a particular set of historical circumstances and obtained only within the terms of a discourse' is *oversocialised*. Owen refers to the work of Hamer and Copeland (1999: 163), which shows how genes influence our sexual desire, how often we have sex etc and, 'help make us receptive to the social interactions and signs of mutual attraction that we feel instinctively'. We have examined the genetic evidence pertaining to how the D4 dopamine receptor gene influences male sexual behaviour, from the work of Hamer and Copeland (ibid) in the section of Chapter Three previously referred to. There is thus no need to rehearse the argument here. In the case of Foucault's (1980b) work, his belief that sexuality is 'controlled' through definition and regulation (rather than by the 'impression' suggested by Adrienne Rich's (1981) feminist analysis of heterosexuality), in particular via the creation of sexual categories such as 'heterosexuality' and so on is open to question. In the case of female sexuality, Foucault (ibid) argues that it is 'controlled' not by the denial or 'silencing' favoured in Walby's (1990) feminist account, but by *constant referral*. Foucault suggests that the 'history of sexuality' is in effect a history of shifting forms of control/regulation. In the case of the last hundred years or so, the shift has been away from the church's moral regulation towards an increased regulation through education, medicine, psychology, law, social work, social policy etc. 'Sex', for Foucault (ibid) is not a biological 'entity' but rather an *idea*, a concept which is always specific to certain cultures and cultural and historical periods. Sexuality for Foucault is produced through *discourses on sexuality*, which shape human sexual values and beliefs. This, as Owen (2006a, 2006b, 2007a) suggests, appears to be an *oversocialised* perspective which denies the notion of a biological sexual drive altogether. As Ridley (1999: 149) shows, 'a gene on chromosome 10 called CYP17' directly affects sexuality in human beings. The gene manufacturs an enzyme which enables the human body to convert cholesterol into cortisol, testosterone and oestradiol, and without the enzyme, the pathway is blocked and the only hormones that can be made from cholesterol are progesterone and corticosterone. People lacking a 'working copy' of this gene 'cannot make other sex hormones so they fail to go through puberty; if genetically male, they look like girls' (ibid).

Owen (2006a) has identified Gagnon and Simon's (1973) symbolic interactionism as *oversocialised*. The authors suggest a very radical form of social constructivism in which there is (as is the case with Foucault (1980b) no sexual drive in the 'natural' sense that it is said all human beings possess to varying extents as part of biological make-up. The authors famously argue for a model of human sexual drive purely as social construction. Not only do we *learn* what 'sex' means, and who/what is sexually arousing to us, but we also learn to want sex. Though the authors acknowledge that the body has a repertoire of biological 'gratifications' (ie, the capacity for orgasm), it does not automatically follow that we would want to engage with them. Certain 'gratifications' will be selected as 'sexual' via the learning of 'sexual scripts'. For example, a particular experience would not be repeatedly sought after unless there was the presence of a 'meaningful' script. From Gagnon and Simon's *oversocialised* perspective, which denies a biological component to *sexual drive* (different from the 'gratifications') socialisation is not about learning to control innate sexual desires in order that they are expressed in socially acceptable ways, but rather the learning of 'sexual scripts' of some complexity which serve to specify circumstances which will elicit sexual desire and 'make' the person wish to engage in certain acts with his/her body. Thus, for the authors, sexual drive is, as it is for Foucault (1980b), a 'learnt social goal'.

The *oversocialised gaze* is an important meta-concept of the framework, and most definitely an illegitimate form of theoretical reasoning, or 'cardinal sin', which we should avoid when engaging in the study of human biotechnology and or related areas pertaining to bio-social issues, especially attempts to 'bridge' the social and biological sciences. However, as we saw in Chapter Four (in the section entitled, 'Biomedical Discourse'), it is equally important to avoid accounts which are *under-socialised.* As we saw in the aforementioned section of the Chapter, the work of Powell and Owen (2005) has suggested that we require a 'balanced' approach to analysis which acknowledges the role of 'the social' alongside the biological in 'biomedical' matters. This is in line with the approach favoured here, which acknowledges a *mutuality* between genes and environment in terms of causality. Following Mouzelis (1991), Powell and Owen (ibid: 28) suggest that 'sociological' approaches, as opposed to rigid, reductionist 'biomedical' approaches are required especially in analysis of the relationship between *macro* and *micro* tiers and links, and between temporal spatialities or contingencies between *agency* and *structure*. Having examined the explanatory potential of the *oversocialised gaze* as a meta-concept of the 'new' framework, we must next look at the concept of the *biological variable*.

THE BIOLOGICAL VARIABLE

At the beginning of this Chapter (in the section entitled, 'The Meta-theoretical Framework'] which essentially involves a codification of the framework/meta-concepts), the *biological variable* is referred to as a 'new' term. The term, or meta-concept, is 'new' in the sense that it is *not* employed by Sibeon. We should regard the meta-concept as referring to the evidence for an, at least, in part *biological* basis for some human behavior. Its use arises out of a respectful critique of Sibeon's original anti-reductionism on the grounds that it neglects biological or part biological causality in explication of social 'reality' (Owen, 2006a, 2006b, 2007a, 2007b; Owen and Powell [2006]; and Powell and Owen, 2005). As was made clear in the previous sections on other 'new' terms (*'the oversocialised gaze'* and *'genetic fatalism')*, the evidence *for* the '*biological variable*' comes from the literature of evolutionary psychology and behavioural genetics in the work of Ridley (1999, 2003), Dunbar (1996),

Hamer and Copeland (1999) and others. The evidence for genetic/partial genetic causality in relation to selected behaviours such as *sexuality* (Hamer and Copeland, ibid; Owen, 2006a; Bogaert and Fisher, 1995); *language* (Hamer and Copeland, ibid; Dunbar, ibid; Enard et al, 2002; Pinker, 1994) *reactions to stress* (Ridley, 2003; Pollack, 1988) have been discussed in some depth in Chapter Three and variously in the sections on 'the *oversocialised gaze'* and '*genetic fatalism*'.

It is the contention here that theoretical analysis, particularly that pertaining to the study of human biotechnology and related areas in which links are attempted between 'the social' and biology (Benton, 1991, 1994; Williams, 1998, 2003; Bury, 1997; Newton, 2003), it is *essential* to recognise that, 'genes do influence behaviour' (Ridley, 199: 306) and that the harshly 'environmentalist' accounts of Foucault (1980b) and Gagnon and Simon (1973) in which sexuality, for example, is completely *learned* behaviour are no longer plausible hypotheses in the age of the Human Genome Project. Wilkie's (1994: 171-72) treatment of the 'moral consequences of molecular biology' contains the expression of unease at the possibility that the Project will 'point up differences between individual humans at a genetic level'. The author suggests that if such genetic knowledge is 'not handled properly and seen in its proper, biological context', it may lead to the generation of information which enables 'new grounds of discrimination'. From the point of view of 'the Project's very existence', we may take an increasingly ''atomistic' view of human beings and indeed of life itself' (ibid). Wilkie's (ibid) greatest fear is that the advances in human biotechnology may lead to a tendency to 'define ourselves in genetic terms and neglect the rest'. The author draws attention to the possibility of the risk that, 'we may all become reductionists reducing our lives to their supposedly fundamental components', missing the holistic 'complexity and richness of life in its entirety'. Arguably, Wilkie's fears are misplaced. There is no reason to suppose that such '*geneticization*' (also referred to by Opitz, 2000) should ever come about. As we have seen in this Chapter's section on '*Genetic Fatalism'*, Schaffner makes it clear that the incompleteness of any *exclusively* genetic basis for human behaviour is taken as a general premise within the scientific community. The *'biological variable'* should be recognised as playing a part in causality, but equally there is a role for the environment (part of the 'complexity' and 'richness' referred to by Wilkie) which is 'massively important' (Ridley, ibid) in the sense that genes are 'switched on' by social cues. Acknowledging a genetic component to causality does not have to entail the *genetic fatalism* engaged in by Wilkie. The three authors of *Not in our genes* (referred to in Chapter Three, in the section entitled 'Building Bridges', pertaining to attempts to create theoretical links between 'the social' and biology), Steven Rose, Leon Kamin and Richard Lewontin made the point that 'biological determinists' believe the credo "you can't change human nature" to be the beginning and end of the human condition. However, as Ridley (1999: 307) makes clear, 'this equation-determinism equals fatalism – is so well understood to be a fallacy that it is hard to find the straw men that the three critics indict'.

In essence, the evidence (in terms of selected examples) for biological variables playing a part in human behaviour has been explored in Chapter Three, and the source of the insights (largely *evolutionary psychological)* has been examined with reference to the staunch defence of such ideas by Curry (2003); Dawkins (1982; 1985); Burns (1994); Daly and Williams (1998); Tooby and De Vore (1987); Cosmides and Tooby (1997), Ridley (1999, 2003) etc, and criticisms by those such as Rose and Rose (2000); David (2002); Rose (2000); Rose, Lewontin et al (1984). It will be recalled that the decision to incorporate the idea of a *biological* basis (which does not imply *fatalism* and acknowledges the *plasticity* and *mutuality* embracing genes and environment/social cues for human behaviour] was made at the end of the third Chapter, after reviewing evidence pertaining to genetic causality/partial genetic causality in the cases of selected examples of human behaviour. Next, we need to examine the explanatory potential of the meta-concept *power* for the study of human biotechnology and or related biosocial areas.

Power

At the beginning of the Chapter (in the section entitled, 'The Meta-theoretical Framework') it is emphasised that the 'new' framework utilises a modified notion of Foucauldian *power* favoured by Sibeon (2004). The author, in a similar position to Derek Layder (2004), acknowledges the *multiple* nature of power. Sibeon's (ibid: 145) view is that, 'power exists in more than one form' – these being *objective, structural* (including *systemic*) forms of power and *agentic* power. The latter meta-concept, *agentic power* refers to what Sibeon (ibid) considers to be the ability of agents to 'shape events in a preferred direction', and this capacity can be both relational and variable. This modified notion of Foucauldian power has also been employed by Owen (2005) and by Powell and Owen (2005) in the field of social gerontology/sociological theory. The latter authors make the point that we should reject, 'the reductionist, methodological collectivist and macro to micro theories of bio-medicine' because they effectively 'view power as an outcome of their causal powers as top down: in any power the gains of one side (bio-medical mode) dominate the other (older people)' (Powell and Owen, ibid: 27).

In Chapter Two, we explored the idea of power in relation to the (arguably) greatest attempt to accelerate bio-medical research and the development of human biotechnology in terms of global scope and complexity – the Human Genome Project (see section entitled, 'The Human Genome Project and Foucauldian Power'). Here, it was argued that the meta-concept of *power* (modified, Foucauldian power – modified in the sense that we must recognise the dialectical relationship between *agentic/systemic* forms; the *relational contingent* and *emergent* dimensions; and that power can be stored) has good explanatory potential in relation to the Genome Project. As Owen (2006b: 909) suggests, we may view the Human Genome Project in terms of power, 'as an example of Sibeon's (1996: 52) 'emergent' power'. We could analyse the vast bio-medical Project, drawing upon Callon and Latour's (1981) theories (though rejecting their idea that power cannot be stored in positions/roles etc), and Foucauldian perspectives and conclude that power is not a fixed capacity, but rather is at least, partly *emergent*, in the sense of being an *outcome* of social interaction. Drawing upon the work of John Law (1986: 5) we could consider power as being, in part, an 'effect' rather than a cause of strategic success achieved by social actors in specific situations. The 'public' version of the Genome Project, it was argued in Chapter Two, must be examined in terms of how power is created and maintained. Therefore, in Foucauldian terms, we need to examine the 'genealogy' of existing relations, how they have 'emerged' and the discourses they may reflect. However, we must repudiate the anti-foundational *relativism* inherent in Foucauldian analysis. Is the Human Genome Project to be regarded as an actor? We need to consider Sibeon's (1996: 57) view concerning the two main (related) points pertaining to social actors when answering that question. For the author, 'they have an ontological status' in the sense that they are able to formulate and act upon decisions using agency, and secondly 'this may be an intermittent status' that does not apply in every circumstance' (ibid). In the case of the Genome Project, as Owen (2006b: 911] makes clear, 'it is an entity that is *not* an actor. To suggest otherwise would be to engage in *reification*'. In some cases, 'the government of a group of 'nation states'' may decide that it is desirable to create a project such as the Human Genome Project that is 'empowered' (Sibeon, ibid: 57) to take decisions/act on the collective behalf of its member states. Here, as was emphasised in Chapter Two, the Genome Project is more akin to a *'supra-national organisation'* (Owen, ibid). It is essential, if we are to adhere firmly to the 'rules' of the new framework, to keep firmly in mind Hindess's (1986: 115) useful formulation of social action; 'a locus of decision and action where the action is in some sense a consequence of the actor's decisions'.

Owen (ibid) makes the point that it might prove fruitful to drawn upon Bennington and Harvey's (1994) concept of transnational and inter-regional policy networks composed of governmental/non-governmental actors and 'apply' it to the public HGP as a whole. In doing so, we could identify the International Human Genome Sequencing Consortium (HGSC) as a '*policy network*'. It is composed of more than 2000 scientists at 20 institutions in six countries, and functions to develop the human biotechnology required to characterise human and other genomes, and to address the ethical/social/legal implications of the 'new' information. As Owen (ibid) suggests, we need to acknowledge the multiplicity of forms that *power* takes in relation to the Genome Project. This includes *systemic* power 'stored', '*contra* Callon and Latour (1981), in discourses pertaining to the Project, the twenty or so institutions that make up the Human Genome Sequencing Consortium (HGSC), the social positions/roles of actors etc and *agentic* power which refers to a capacity of actors'.

In Chapter Four, we examined six techniques of human biotechnology that, in principle, according to Lynn (2001) have been utlised to promote *eugenics* or selective breeding. These are artificial insemination by donor (AID); egg donation; pre-natal diagnosis of fetuses with genetic diseases/disorders and termination of the pregnancies where these are identified; embryo selection which consists of the 'ascertainment of the genetic characteristics of embryos grown in vitro and the selection for implantation of those with genetically desirable characteristics' (ibid: 245); cloning, consisting of the production of genetically identical copies of individuals; and genetic engineering, in the sense of 'the insertion of new genes' (ibid). As emphasised in the Chapter (the section entitled, 'Further Ethical Concerns and Human Biotechnology'), human biotechnology has given birth to a new academic discipline in its own right – *Bioethics*, with its own journals in which ethical questions are addressed and debated. It is *not* the intention here to cover the literature on bioethics, and that would indeed be beyond the scope and intention of the book. It is (or rather was) necessary to cover the major ethical concerns pertaining to the six selected examples (above). Arguably, to reiterate a point made in Chapter Four, one could not credibly cover the subject of human biotechnology without discussing ethical concerns (this does *not* mean 'taking sides' on moral grounds) because the two are linked so strongly in the pubic perception (David and Kirkhope, 2005; Wilkie, 1994). As Lynn (2001: 258) has suggested, although techniques such as pre-natal screening, IVF etc have a 'eugenic impact' and potential, the ethical problems are, 'not the same as those raised by classical eugenics'. As Lynn (ibid) shows, in 'classical' eugenics, the state, which in the metatheoretical framework employed here is *not* regarded as an actor or agent because it cannot formulate/act upon decisions, 'can legitimately interfere in the population's reproduction'. Hitler's Third Reich is the most notorious example of this. In Western democracies, the state does not promote the use of human biotechnology to influence the population's fertility and promote eugenics. As Lynn has suggested, 'the techniques' of reproduction are utilised by autonomous *individuals* (that we must regard as *actors*) to enable childbirth. Here, the philosopher Kitcher's (1996) view that we are living in an age of '*laissez faire*' eugenics may apply. This is the idea, referred to in the section of Chapter Four mentioned above, that every individual actor has an opportunity to be 'their own' eugenicist, taking advantage of available genetic tests (for example, pre-natal screening) in order to make the selective reproductive decisions they view as 'correct'. This then, if analysed in keeping with the modified conception of Foucauldian power favoured in the 'new' framework, is a case of *agentic* power, in the sense that it refers to the ability of individual social agents to 'shape events in a preferred direction' (Sibeon, 2004: 145). By utilising 'techniques' of human biotechnology, individual actors are taking control of reproductive events. There are, of course, moral issues here which we cannot pursue further because that would be beyond the scope of the book.

Ridley (1999: 298) makes the point that Western democracies have rejected government eugenics programmes, 'merely to fall into the trap of allowing private eugenics'. He refers to the 'pressures' parents are subject to with regard to adopting 'voluntary eugenics' from professionals such as doctors, health insurance companies and 'from the culture at large' (ibid). The point is made that 'stories abound' of women 'as late as the 1970s' being 'cajoled' by doctors into sterilisation 'because they carried a gene for a genetic disease' (ibid). In the cases Ridley refers to, we might 'apply' the idea of Sibeon's (2004: 136) *systemic* power, in the sense that doctors and other professionals (aside from being individual social actors capable of 'possessing' power) occupy *roles* and *positions* within which, *contra* Foucault (1991) and Callon and Latour (1981), 'certain elements of power can be stored in positions/roles, social institutions and social systems'. As Sibeon (ibid) has suggested, Foucault and Callon and Latour, 'tend to push their relational and processual conceptions of power to the point of denying that power can be 'stored' in roles and in social systems and networks of social relations'. Here, we have adopted Sibeon's (ibid) synthesis of a combination of Foucauldian and other *relational* concepts of power, with a *systemic* understanding of power. As Sibeon (ibid) suggests, the synthetic conception of power leads to the idea that 'agentic power nearly always has a relational, contingent and emergent dimension', which is arguably why the position of what Sibeon calls 'top dogs' in any institutional sphere is sometimes 'precariously sustained'.

We therefore would regard as legitimate the idea that an actor, such as a doctor in the cases of patients being 'cajoled' into 'sterilisation' referred to by Ridley (ibid), 'has' power (which invokes the idea of 'storage') as long as we do as Sibeon (ibid) suggests, and recognise 'how that power is constituted and reproduced relationally'. The author makes a reference to the work of John Law (1991b) here in relation to the latter point. As Sibeon (ibid) has made clear, there is good evidence to regard power as, 'partly preconstituted and stored' (in roles etc), and 'partly relational, emergent and contingent, with the extent to which power is 'systemic' or 'relational' being an empirical variable' that may vary situationally. This leads to the idea that there are multiple forms of power like agentic, systemic and relational ones previously mentioned. In the case of power's 'multidimensionality' (Sibeon, ibid: 137), we must consider the possibility that there is a 'multi-level aspect' to this phenomenon. In other words, there could be 'hybrid', 'more variegated' forms of power than systemic-relational distinctions might imply. As Sibeon (ibid) suggests....

Agentic power in some circumstances has a largely systemic source deriving, say, from position/role. In other circumstances agentic power may be of a relatively contingent, emergent kind that emerges during and as an effect of social interaction at the micro or mezo (for example, inter-organisational) levels of social process. Such power may interact with systemic or positional/role power of the type that is 'stored' in discourses, social institutions, and social systems/social networks. There tends, in other words, to be a two-way (dialectical) relation between systemic and relational forms of power, with each to some extent conditioning the other.

Discourse may then, in a sense, be seen to 'embody' power, for example, the discourse around/about 'techniques' of human biotechnology and reproduction such as IVF and pre-natal screening. The discourse 'stores' power which relates to multiple, countless decisions taken by professionals and administrative actors that, in turn, 'shape' the structure within which the professionals such as doctors, scientists etc operate and influence agency/decision-making. However, we must apply great caution here. Discourses themselves are *not* actors. *Contra* Foucault, discourses are (in the sense that they are patterned ways of thinking and behaving), 'a form of material that must be mobilised by *actors*' before the discourse can be regarded as having, 'any social consequences or effects' (Sibeon, ibid: 72).

Of great relevance here is Derek Layder's (1997, 2004) multi-dimensional conception of power. Sibeon (2004: 187) notes how Layder's convincing attempt to 'construct a multi-level research strategy' has 'a certain amount of affinity' with his own. Layder's stratified social ontology is, as Sibeon (ibid) notes, 'a modification and elaboration' of the *distinction* between 'micro-, mezo- and macro- domains'. Layder (1997), as we have previously noted in Chapter Two (in the section devoted to the author, 'Derek Layder') identifies four *social domains* or 'layerings' as he has sometimes called them. These four *domains* have been defined and discussed in the aforementioned section of Chapter Two. However, to briefly recap, they may be defined as follows. *Psychobiography* refers to largely unique, asocial components of self and behaviour. *Situated activity* refers to face-to-face interaction and intersubjectivity in situations of co-presence. *Social settings* are the locations in which situated activity occurs. *Contextual resources* relate to divisions of class, gender and ethnicity, consisting of macro-distributions and ownership of resources. Each domain has relative autonomy but they may overlap and influence each other. As Sibeon (ibid) makes clear, any connections developing between them must be viewed as 'loose', lacking direct links between the micro, mezo and macro tiers of his own research strategy, or in Layder's terms, the four domains (psychobiography, situated activity, social settings and contextual resources). Power for Layder (1997: 13, 250-1) concerns many forms and is 'objective' or external to individual subjectivity, embodied in discourse (which Foucault, [1980b] acknowledges, and Sibeon,[2004] appears to agree with in the sense that power can be 'stored'), but can also be 'subjective'/intersubjective. Sibeon (ibid: 184) interprets Layder's (ibid) conception of power as having 'relatively independent, distinct and domain-specific modes of existence'. Power 'wends its way across the micro-, mezo- and macro- domains that constitute social life' (ibid). Power too, like the four domains Layder identifies, may be interpenetrative and overlapping, with more or less loose or indirect links between the different forms that power assumes either in terms of the asocial aspects of the person, the face-to face interaction, the locations in which interaction/intersubjectivity occurs, or the divisions of class, gender, ethnicity etc.

Previously in this section we saw how Owen (2006b) has argued that the Human Genome Project may be viewed as a *'supra-national organisation'* rather than an actor in the sense that it does not 'possess' the agency to be able to formulate and act upon decisions. The point was also made, regarding power, that we could regard the Genome Project as an example of 'emergent power', in the sense that power is not 'fixed' but at least partly the outcome of social interaction (Owen, ibid: 909). We also considered the importance of acknowledging the multiplicity of forms that power takes, such as *systemic, relational* and *agentic* forms. Power, as Owen (ibid) emphasises may be 'stored' in discourses pertaining to the Genome Project and in the roles/positions of actors in the institutions that make up the Human Genome Sequencing Consortium (HGSC), in the form of *agentic* power. If we examine the Human Genome Project in terms of Layder's *social domains* theory, we might view power as 'stored' in the domains of the individual *psychobiography* of key actors; in the face-to face interaction and intersubjectivity involving the co-presence of actors in their various roles and positions (scientists, administrators, politicians etc); in the *social settings* such as the laboratories of the twenty or so institutions of the HGSC in which *situated activity* occurs; and in the *contextual resources* pertaining to divisions of class, gender and race among actors. Power would also be seen as 'embedded' within the discourse on the Genome Project. Regarding the latter points, it is sobering to consider what Bucchi (2004: 132) calls 'the huge impact of the Genome Project on the public imagination'. Regarding the discourse on the Project (within which power is 'stored' or 'embedded'), a wealth of striking metaphors have been utilised to describe the huge bio-medical undertaking, such as 'the book of life', 'map', 'cartography', 'the instruction manual for our species' etc, and these highlight its public resonance (ibid). As Bucchi (ibid) suggests, 'considerations of political and social expediency' tended to hamper its acceptance by scientists, but thereafter scientists and philosophers were included in the Project to study its implications for society and ethics. On the other hand, as Bucchi (ibid) acknowledges, some (perhaps the maverick Craig Venter of Celera, the rival 'private' Genome Project discussed in Chapter Four, in the section entitled, 'The Human Genome Project', is a case in point in terms of media publicity) have 'widely exploited its visibility'. Arguably, the associations in the public domain, in what Layder (1997) would term *social settings* too, between genetics and the treatment of disease has ensured that the Project reached 'completion' in the form of a finished sequence of the human genome in 2001, despite 'the opposition of numerous specialists' (Bucchi, ibid). The discourse on the Project is huge, and, for example, the *New York Times* featured '1,069 articles on the human genome between 1996 and 2001' and the peak in coverage by the newspaper coincided with the political interventions at the highest level (Prime Minister Blair and the then President Clinton) in 2000 and the promise by both the 'public' Human Genome Project and Celera that the enterprise was near completion, rather than the final joint announcements that the genome had been 'mapped' (Bucchi, ibid: 133).

Having demonstrated the explanatory potential of the 'new' framework's modified notion of Foucauldian power, we next move on to the usefulness of the concept of *dualism of structure* in studies of human biotechnology and related areas involving links between the social and biological sciences.

DUALISM

In Chapter Two (the section entitled, 'Agency and Social actors') we discussed the Giddensian rejection of dualism in favour of a duality of structure. Later in the same Chapter (in the section entitled, 'Archer, Layder and Mouzelis: Post-postmodern Theorists') we discussed the approaches to the question of 'dualism or duality?' in relation to the theoretical and research strategies of the aforementioned major sociological thinkers. Chapter Three saw further and more involved discussion of the question, this time pertaining specifically to biosocial issues such as the attempts to 'build bridges' between the life sciences and 'the social' by authors of the 'embodied school' or 'sociology of the body' such as Williams (2003), Benton (2003), Newton (2003) and others in 'Duality Versus Dualism Revisited'. We saw that most 'material-corporeal' sociologists appear to be very much united in a desire to, 'challenge dualism on all fronts' (Williams, ibid: 556). Benton (2003: 292), it will be recalled, argues for an approach which would 'dissolve dualistic oppositions' (nature/society).

To recap, as we have seen at the beginning of this Chapter, *dualism* is favoured because, it is felt that in social analysis, *agency-structure* and *micro-macro* should be employed as dualisms that refer to distinct, relatively autonomous phenomena. We seek here to avoid what the 'embodied' writer Shilling (1993) argues for in his analysis of the human body as 'simultaneously' biological and social. Here, Shilling (ibid) has arguably collapsed both terms into an amalgamated whole within which elements cannot be separated. This is a similar position to Giddensian *structuration*, and the Foucauldian (1972, 1980a, 1980b) tendency to compact *agency,* and *structure* together. As Owen (2006a) has suggested, if we collapse distinctions between either *agency-structure* or *biology – the social*, it results in what Archer (1995) calls 'central conflation'. This emphasis upon dualism, it should be emphasised, reflects the influence of the work of Roger Sibeon and Derek Layder upon the construction of the framework. Both authors favour *dualism* rather than the elisionist tendencies of Giddens. Sibeon (2004: 98) makes the point that a 'failure to acknowledge temporality' will result in an 'inability to conceptually separate agency and social structure in a way that might allow us to study links between the two and allow us to assess the reflective influence of each in any given social context'. This is arguably correct, and applies equally to *biology* and *'the social'*.

It will be recalled that we favour here a conception of the relationship between genes and environment, in relation to causality, that Ridley (1999, 2003) calls 'Nature via Nurture'. In other words, we acknowledge that there is a 'feedback loop' that embraces genes and environment, a mutuality which means that genes predetermine the broad structure of the human brain and absorb formative experiences, and react to environmental 'social cues'. This does not mean, however, that we cannot analyse genes and environment, biology and the social separately. Maintaining *dualism* as a meta-concept enables us to do so. In Chapter Four (in the section entitled, 'Bio-Medical Discourse'), we discussed the recent work of Powell and Owen (2005), who have drawn attention to some possible dangers associated with a misuse of dualism. As the authors suggests, we must guard against 'mind-body' dualisms becoming 'the location of regimen and control' in positivistic, methodological searches for objective truths. As is made clear in the section of Chapter Four mentioned above, we need to guard against the misuse of dualism in the form of bio-medical totalising analyses which emphasise biology in causation, leaving out 'the social' altogether. As we saw, it is regarded as just as illegitimate to posit *under-socialised* perspectives as it is to engage in *oversocialised* analysis. We need to acknowledge that genes and environment *both* play a role in human behaviours and *dualism*, if correctly employed is a useful concept which enables the separation of elements like agency, structure, biology, 'the social' etc, enabling us to examine the links between them and to assess the relative influence of each element in any social context.

AGENCY-STRUCTURE, MICRO-MACRO, TIME-SPACE

To recap, all these terms are defined at the beginning of the Chapter (in the codification of the framework that is entitled, 'The Meta-theoretical framework'). However, briefly the terms or meta-concepts may be defined as follows. *Agency* is defined in strictly non-reified terms, in which actors are conceptualised as entities capable of formulating and acting upon decisions. *Social structure* is defined as 'social conditions', not a unitary phenomenon but rather contingently reproduced. *Micro-macro* refers to the units/scales of analyses concerned 'with the investigation of varying temporal and spatial extensions of the social' (Sibeon, 2004: 57). *Time-space* refers to relatively neglected concerns within social theory pertaining to temporality and spatiality.

As previously stated, the emphasis in the 'new' framework is upon a concept of *agency* which is non-reified. That is to say that it avoids at all cost what Sibeon (2004) identifies as *reification*; the attribution of agency to an entity incapable of formulating and acting upon decisions. However, like Barry Hindess (1986a), Roger Sibeon argues that there are actually *two* main types of actors; *individual human actors* and *social actors* (the latter are termed 'supra-individuals' by Harre, 1981: 150-2), such as organisations, families, committees, central government departments, professional associations etc. As Sibeon (2004: 119) puts it, the decisions of such *social actors*, 'shape much of the social, economic, and political terrain' in today's society. They are not mere aggregations of the decisions of individual human actors. This relates to the concept of '*emergence*' (ibid: 120), in the sense of emergent properties arising within social structure. However, the form of *agency* which *social actors* (such as a political party) are capable of, in terms of decisions and actions, is different to that of individual human actors and it is essential not to confuse the two. Thus, in relation to the Human Genome Project, the US Department of Energy (DOE) which originally explored the feasibility of such a mammoth bio-medical undertaking may be regarded as a '*social actor*' or a '*supra-individual*', depending upon whether we draw upon Hindess (ibid) and Sibeon (ibid) or Harre (ibid) respectively. It cannot, however, be regarded as an individual actor on par with a sentient human being capable of the kind of *agency* human individuals require to formulate/act upon decisions. To imply or suggest so would be to engage in the 'cardinal sin' or *reification*. When Owen (2006b: 911) refers to the Genome Project as a 'supra-national organisation' rather than an 'actor' (in Chapter Two's section on 'The Human Genome Project and Foucauldian Power', in Chapter Five's section on 'The Human Genome Project', and this Chapter's section on 'Power') it should be emphasised that Owen means *actor* in the sense of an *individual actor*. As to whether the Human Genome Project 'qualifies' as a *social actor* alongside the, 'organisations in the state, private or voluntary sectors, including interest groups, political parties' etc identified by Sibeon (ibid: 119) is a question in point. Sibeon's conception of a social actor includes central government departments like the DOE above, but does not mention such phenomena as international research programmes like the Project. Certainly, within the infrastructure of the Project there were 'organisations' such as the Human Genome Sequencing Consortium consisting of many institutions, and the Project had the 'institutionalised mechanisms of resource attraction and allocation' and 'formal networks of positions/roles' etc which Sibeon (ibid: 120) refers to as characteristic of *social actors*. However, perhaps we should proceed with caution here. The author Bucchi (2004: 129) compares the rival private company Celera which competed in the 'race' to sequence the human genome with the 'public' Human Genome Project, an international consortium', in operation for more than a decade, and 'financed with billions of dollars by some of the richest industrialised countries'. We have seen, in Chapter Four (in the section on 'The Human Genome Project') how Bucchi (ibid: 134) considers that the Genome Project 'exemplifies a number of ongoing patterns of change in contemporary science'. These 'changes' refer to the emergence of a 'new' 'Big Science' in the form of collective endeavours undertaken by huge, complex organisations; the emergence of a 'post-academic' science in which disciplines intersect, social accountability is an important issue, and the actors involved are hetereogenous (Bucchi, ibid). Also referred to by the author is the emergence of a 'new' 'academic-industrial-governmental' complex of which the Human Genome Project is a 'product' (ibid). Perhaps some would regard the Project in itself as a *social actor*, but here we will be cautious and reiterate that whilst the Project is *not* an actor in the sense of possessing agency of the sort associated with individual human actors, and it may be seen as a '*supra-national organisation'* (Owen, 2006b: 911), there is no mention of international research programmes in the definitions of *social actors* contained in the work of Sibeon (2004).

In an earlier section of this Chapter devoted to the 'cardinal sin' of *reification*, we discussed the theories of *Actor-network* and *posthumanism*, which are greatly at odds with the non-reified conception of agency favoured in the 'new' framework. The point was made that we must reject both kinds of accounts. *Material* and *non-human agency* which attributes causal powers to physical objects (Pickering, 2001) such as stones must be repudiated.

Micro-macro as a meta-concept is applied in the 'new' framework in the sense that Sibeon (2004: 46) intended; in other words, 'micro-macro should not refer to differences in the absolute "size" of social phenomena, but rather, to *relative* differences, in size and to a relational conception of scale associated with the concept of 'emergence''. *Emergence* can be found at 'micro-, mezo- and macro- levels of social process' (ibid: 76). Sibeon (ibid) provides an illustratory example of emergence in the case of a, 'formal committee or organisation in the public sector' which is 'likely to have institutionalised mechanisms of legitimisation, resource-attraction and decision-making' and 'causal powers' that 'cannot be reduced to the level of its individual members' (Holzner, 1978; Clegg, 1989: 187-8). The point is also made by Sibeon (ibid), that 'informal groups also have emergent properties' and these can take the form of 'informal systems of rules'. In a *macro* sense, the Genome Project can be viewed as 'Big Science' in which 'venture capitalists set the research agenda, researchers act as entrepreneurs, and politicians confusedly mediate between the two sides' (Bucchi, 2004: 135). On a *micro* level, the Genome Project is embedded within the relatively small-scale, 'heterogenous social networks' (ibid: 134) comprising journalists, politicians, businessmen, and even young computer hackers. As Merriden (2001) shows, 'Napster', peer-to-peer technology for sharing resources was one model used by human genome researchers to coordinate various genomic sequencing centers.

As Sibeon makes clear, the *spatial* dimension of society has not received even, as much as the (already neglected) *temporal* dimensions. *'Locale'* is posited by Sibeon (ibid) as a 'bridging concept' to span micro-, mezo-, and macro- levels of the social. In his view, 'some local events may be specific to the location in question, others may be connected to the macro-social order' (ibid). Therefore, we are advised to 'examine the internal properties of local sites as well as the emergent properties of temporal and spatial linkages across sites'. We might bear this in mind, and focus upon the *internal aspects* of local sites when studying the Genome Project. The completion of the draft sequence of the human genome was the culmination of 'more than a decade of work involving 20 sequencing centers in six countries' (Dennis and Gallagher 2001: 36). Sibeon (ibid: 167) stresses the concepts of *'networks* and *'materials'* in relation to the investigation of the properties of *time-space*.

Networks/systems are 'more or less patterned relations between actors between, for example, positions/roles' and it is partly though not exclusively, 'within networks' that materials 'flow' (ibid). Here the term *'material diffusion'* applies. The concept of materials refers in Sibeon's (2004) work to discourses, social practices and typifications where 'these phenomena are inspected in terms of their tendency to 'travel' across spatio-temporal contexts' (ibid). In terms of explanatory potential with regard to the study of human biotechnology (and or related areas such as the links between life sciences/'the social'), these concepts can be employed for analytical purposes. If we are to study, for selected example, the six techniques of human biotechnology identified by Lynn (2001) and discussed in some depth with regard to issues pertaining to their perceived social and ethical (especially eugenic) implications in Chapter Four, we could do so with regard to *materials* and *material diffusion* processes of the kind identified by Sibeon (ibid). The six examples of human biotechnology were artificial insemination by donor (AID); egg donation; prenatal diagnosis of fetuses with genetic diseases and disorders, and termination of the pregnancies where these are identified; embryo selection; genetic engineering and cloning. Arguably relevant *materials* to examine in the investigation/study of these selected examples of human biotechnology for analytical purposes can be regarded as consisting of existing laws, for example on cloning and sex selection; norms and conventions associated with biotechnological procedures; management, training and professionalism; conflicting moral-political conceptions around the nature and purpose of reproductive human biotechnology etc. We might regard the *discourse* pertaining to these examples of biotechnology as existing spatially and temporally within and across a variety of *micro* sites characterised by face-to-face interaction, together with *macro* locales such as organisations, hospitals etc. We would then be able to concern ourselves with the detailed analysis of *how* time-space links between relevant (micro, mezo, macro) locales are maintained or changed; the analysis of *micro* (interactional) processes related to the application/handling etc of material *within* the various locales; and the study of actors (individual human actors and or social actors) involved in material diffusion *across* locales across time and space in the social context.

CONCLUSION

We have now examined the arguably useful explanatory potential of the 'new' metatheoretical framework and its meta-concepts in relation to the study of selected examples of human biotechnology and closely related areas in which efforts have been made to some extent to 'build bridges' between the social and life sciences (writers of the 'embodied' school, posthumanism etc), and explore biosocial issues. As we saw earlier in this Chapter (in the section entitled, 'The Meta-theoretical Framework', which is essentially a codification of the 'sensitising device' and its concepts), the framework is capable of contributing both to the 'return to' sociological theory and method, and towards the 'post-postmodern' study of human biotechnology. It also is a contribution towards the building of links between 'the social' and biological science. The 'pay off' will hopefully lie in the framework's explanatory potential in terms of further theoretical development. Rather than being merely 'hot house debate', it is contended that the framework's usefulness lies in its avoidance of anti-foundational *relativism; oversocialised* (harshly 'environmentalist') accounts; the tendency towards *genetic fatalism* (the equation of genetic predisposition with inevitability) in the work of writers associated with the sociologies of health and medicine; and its adoption of the 'sociological realism' associated with Sibeon (1996, 1999, 2004, 2007), Layder (1997, 2004, 2007), Mouzelis (1991, 1993, 1995, 2007) and Archer (1988, 1995, 1998).

We have applied the following meta-concepts to the following areas in Chapter Five in order to emphasise and demonstrate explanatory potential. *Reification*, or the illegitimate attribution of *agency* to entities which are *not* actors (ie the state) as a 'cardinal sin' was identified in the work of Wilkie (1994: 187), who makes the claim that institutions 'have a will to live' in relation to his discussion of the implications of the Genome Project. We saw how Cazillis (2001) has identified 'embryo reification' in the use of stem cells in human biotechnological techniques. Actor-network theories such as those of Callon and Latour (1981) and Posthumanism associated with Pickering (1993, 1995a, 1995b) were examined in terms of their conceptions of 'the actor' both of which are greatly at odds with the non-reified conception favoured here. Williams-Jones and Graham (2003) were criticised too for their reified conception of a susceptibility test for hereditary breast cancer. The authors, influenced by actor-network approaches, attempt an ethical analysis of commercial gene testing but, like Callon's (1986) treatment of scallops, they engage in the illegitimate attribution of agency when they claim that the susceptibility 'interacts' and 'shapes' technologies, people and institutions.

We identified *essentialism*, as a 'cardinal sin' and 'form of theorising that in a prioristic fashion presupposes a unity' of social phenomena (Sibeon, 2004: 4) in the work of Buctuanon (2001). The author appears to regard globalisation (in relation to biotechnology) as a homogenous phenomenon. Additionally, her claim that 'biotechnologies tend to agglomerate in developed economies, leaving to pure markets and limiting state involvement', (ibid: 26) leave us no wiser as to whether she refers to gene-enhancement products, reproductive technology etc, with 'techniques' and 'technologies' never specified. A tendency towards *essentialism* was identified in the work of the feminist theorists such as Hanmer (1993). Hanmer appears to refer to women as if they are an essentialist, monolithic block in terms of women *per se* (with no distinctions based upon their class, gender, race etc) 'having no option about whether to have babies' (ibid: 232). Her claim (Hanmer, ibid: 234) that 'men as a class' are 'taking control of' women's reproduction activities was also identified as *essentialist* thinking of the kind cautioned against in the 'new' framework.

Functional teleology as a 'cardinal sin', and meaning, 'an invalid form of analysis involving attempts to explain the courses of social phenomena in terms of their effects' (Sibeon, 2004: 6) was identified in relation to bio-medical approaches of a Marxian nature which attempt to explain old age as an effect 'or in terms of needs of capitalist social systems' (Powell and Owen, 2005: 28). Hanmer's (1993: 232) critique from a feminist standpoint of 'the new reproductive technologies' was again identified, this time for the *functional teleology* implicit in her argument that such techniques function to uphold patriarchal 'notions of motherhood and femininity via the selection of those who should or should not be mothers'.

Reductionism as a 'cardinal sin' which 'illegitimately attempts to reduce the complexities of social life to a single, unifying principle of explanation' (Hindess, 1986a, 1988 in Sibeon, 2004: 2), was identified in Farganis's (1994: 15-16) work which suggests that gender is the ultimate, primary axis of social life. Buctuanon (2001: 26) is again identified as relying upon a 'classic', *reductionist* model of globalisation in relation to biotechnology, which underplays the 'uneven' and 'overlapping' tendencies of globalisation processes as emphasised by Sibeon (1996: 153). O'Brien's (1981, 1989) feminist accounts of the 'politics' of human reproduction engages, we identified, in the *reductionist* theory that the 'commodification' of life processes is structurally directly related to the 'interests' of capitalism. We also identified possible *reductionist* tendencies in the work of Nelkin (1994), Turner (2004), Hall (2003), Alexander (2003) and West (2003), all of whom, in differing ways *reduce* the complexities of human biotechnologies to forms of 'new' *religion*.

We identified *genetic fatalism*, as a 'cardinal sin' and tendency 'within contemporary social science to equate genetic predisposition with inevitability' (Owen, 2006a: 185) in the work of those such as Rose, Kamin and Lewontin (1984) who appear to have argued that causality in relation to human behaviour lies 'not in the genes'. We have shown how Ridley (1999, 2003) points out that the idea that genetic influences upon behaviour amount to 'slight or non-existent' (as suggested by Rose et al, ibid) is a view, 'no longer tenable' in the light of studies in behavioural genetics which clearly reveal that 'genes do influence behaviour' (Ridley, 1999: 306). For Ridley (ibid), those such as Rose et al (ibid) make a fundamental error in associating genetic predisposition (Ridley uses the phrase 'determinism' too) with inevitability. In his view, this reflects a misinterpretation of the literature of behavioural genetics and evolutionary psychology, and equates accounts which acknowledge genetic causality with fatalism. As Ridley (ibid) acknowledges, with regard to causality, 'environment is still massively important', and we should think not in terms of 'nature versus nurture', but rather 'nature *via* nurture'. In other words, we need to acknowledge the *mutuality* and *plasticity* of genes and environment. The work of Schaffer (2000) was identified as engaging in *genetic fatalism* in the sense that he associates 'genetic determinism' with the inevitable. The view here is that the author need not posit 'genetic-environmental determinism' as an alternative because Ridley (1999: 307) has shown how 'the equation of determinism with fatalism is a fallacy' because of the 'feedback loop' between 'nature' and 'nurture'. *Genetic fatalism* was also identified in the work of Newton (2003) of the 'embodied' school of sociology, specifically his treatment of Benton's (1991:17) work on Cannon (1914). Newton (ibid: 29) appears to dismisses Cannon's (ibid) accounts because they rely upon the suggestion that emotions of anger or rage or fear are associated with 'instincts'. We have previously encountered contrary evidence in Chapter Three from the work of Ridley (2003) and Pollock (1988) for an *instinctive* basis for reactions to stress, alongside evidence from those such as Hamer and Copeland (1999) and Bogaert and Fisher (1995) which suggests that *sexuality* is innate not merely 'learned'; and from Hamer and Copeland (ibid), Dunbar (1996), Enard et al (2002) and Pinker (1994) which suggests the existence of a language *instinct* in Homo-Sapiens. Wilkie's (1994) work is criticised here for associating the view of genes as 'immutable' with 'sociobiological' accounts. As we make clear, Wilkie (ibid) appears oblivious to the reforms from within sociobiology which led to the development of a more 'sophisticated' evolutionary psychology in the form of the work by Tooby and Cosmides (1992) in Barkow et al (1992) which argues for a more flexible picture of causality in which genes play a part, but environment and the underlying mechanisms pertaining to psychology play a part also.

We identify examples of *the oversocialised gaze* as a 'cardinal sin', and tendency within 'mainstream' social science towards harshly 'environmentalist' accounts which are antipathetic towards genetic, or even partially genetic accounts of human behaviour. Giddens' (1993: 57) denial of instincts in human beings 'in the sense of complex patterns of unlearned behaviour' is a case in point. Again, Newton's (2003) work is identified here and his previously referred to rejection of accounts of stress which rely upon genetic/partially genetic explication are regarded as 'oversocialised'. Newton cites only complex social factors in relation to causes of stress. However, as Ridley (1999: 149-50) shows, genes *and* environment play a part. The hypothalamus, 'taking' orders from social cues stimulates the pituitary, which in turn stimulates the adrenal cortex. In a sense, 'nobody' is 'in charge', but the response to stress in Homo-Sapiens depends upon natural selection's intricate, interconnected system in which 'nurture' depends upon genes, and genes require nurture in a form of elegant plasticity and mutuality. Gagnon and Simon (1973) and Foucault (1980b) are both identified as engaging in *oversocialised* accounts of human sexuality. Without going into the evidence offered in Chapter Three (the section on 'Incorporating the Biological Variable') and Chapter Five on the 'cardinal sin' in question too deeply, we consider contrary evidence from Hamer and Copeland (1999) and Ridley (1999) which pertains to the influence of the 'D4' dopamine receptor gene upon male sexual behaviour, and pertains to a gene on chromosome 10 called 'CYP17' which directly affects human sexuality *per se*, respectively. In addition to cautioning against the reliance upon *oversocialised* perspectives which deny biological variables in causality altogether, such as Foucault's (1980b) theory that sexual drive is a learnt social goal, we also warn against *undersocialised* conceptions of the person and causality. Powell and Owen's (2005) work on the 'bio-medical' approaches to human aging are drawn upon here. It is suggested that we strive for a 'balanced' approach to analysis which acknowledge the *joint* role of 'the social' and genes in relation to causality.

The Biological Variable here is intended to refer generally to the evidence for genetic or at least partially genetic bases for human behaviour. The evidence we draw upon in Chapter Five, and consider in depth in Chapter Three, is from the disciplines of behavioural genetics and evolutionary psychology. As previously stated, we considered selected examples of human behaviour, such as reactions to stress, language and sexuality for which there have been posited 'environmentalist' or *oversocialised* explanations and offered contrary evidence from a range of sources across the disciplines of evolutionary psychology, behavioural genetics and biological science such as Ridley (1999, 2003); Hamer and Copeland (1999); Enard et al (2002); Lai (2001); Dunbar (1996); Pinker (1994); Harris (1998); Bogaert and Fisher (1995); Curry (2003); and Cosmides and Tooby (1997) etc. After careful and cautious examination of this evidence, the case was made at the end of Chapter Three (section on 'Incorporating the Biological Variable') for incorporating such cogent evidence for, at least in part, *biological* causality. It is the contention here that Sibeon's (1996, 1999, 2004) original framework has enormous explanatory potential. However, if we are to study subjects such as human biotechnology and the related biosocial questions of linkage or possible linkages between biology and 'the social', we do need to incorporate a recognition of *the biological variable* into analysis.

We applied the modified, Foucauldian notion of *power* to the study of the Human Genome Project and to selected areas of human biotechnology in order to demonstrate the meta-concept's explanatory potential. Here, we decided that power should be regarded as having *multiple* origins, as is found in the work of Sibeon (2004) and Layder (1997, 2004). As Owen (2006a: 193) has put it, Foucault's conception of *power* may be successfully incorporated into analysis, but this must be done 'in a critical and selective fashion, informed by the critique of *agency-structure*, *micro-macro* and other terms offered in the 'new' framework'. If we are to employ Foucauldian insights pertaining to *power*, we must recognise the *dialectical* relationship between *agentic* and *systemic* forms of power; the *relational, contingent* and *emergent* dimensions of power and very importantly, the concept that *contra* Foucault, aspects of power can be 'stored' in positions/roles and in social systems/networks. We identified the Human Genome Project as an example of Sibeon's (1996: 52) 'emergent power' in the sense of being an *outcome* of social interaction, and a *'supra-national organisation'* (Owen, 2006b: 911). The International Human Genome Sequencing Consortium (HGSC), composed of more than 2000 scientists at 20 institutions in 6 countries, drawing upon Bennington and Harvey's (1994) concept of transnational and inter-regional policy networks, was identified as a *'policy network'* (Owen, ibid). *Agentic power*, in the sense that it refers to the ability of individual human actors was identified in the case of individual people who are 'taking control' of reproductive events in the context of utlising new techniques of reproductive human biotechnology. In the words of Kitcher (1996), we are living in an age of '*laissez-faire*' eugenics, in which every individual has the 'power' to be his/her own eugenicist, taking advantage of available biotechnology, such as pre-natal screening, IVF etc, in order to make the selected reproductive decisions they view as 'correct'. We identified cases of *systemic power* in the roles/positions of doctors and other professionals (in the sense of power 'stored' in roles) who Ridley (1999: 298) suggests may 'cajole' women 'into sterilisation' if they carry certain genes for diseases. Drawing upon Derek Layder's (1997) concept of *social domains*, we identified *power* as 'stored' in the psychobiographical domains of individual key actors of the Human Genome Project, in the face-to-face interaction/intersubjectivity of actors in their various roles/positions; in the *social settings* such as laboratories/institutions of the HGSC (Human Genome Sequencing Consortium) in which *situated activity* occurs; and in the *contextual resources* pertaining to divisions of class, race and gender among actors.

Dualism was favoured (as it is by Sibeon (1996, 1999, 2004) and Layder (1997, 2004), as opposed to a *duality of structure*. This was because we seek to avoid the 'central conflation' that Archer (1995) cautions against pertaining to the collapsing of distinctions between *agency* and *structure*. Arguably, we need to avoid elisionist tendencies with regard to biology and 'the social' too. Material corporeal sociologists of 'the embodied' school such as Shilling (1993) argue for a conception of the person which is 'simultaneously biological and social'. Whilst we acknowledge the mutuality between genes and environment (genes can be 'switched on' by social cues Ridley (1999, 2003)), there may be times when it is necessary to analyse both 'parts' separately.

The meta-concepts of *agency-structure, micro-macro* and *time-space* were 'applied' to the study of human biotechnology and related biosocial areas. The distinction was made between *individual human actors* and *social actors*. The latter, such as political parties, 'organisations in the state' etc as identified by Sibeon (2004: 119), are *not* capable of agency in the same way as the former. However, their decisions (collective) can 'shape' much of the social landscape. We identified the US Department of Energy (DOE), which originally explored the feasibility of launching the Genome Project, as a *social actor*. However, it was necessary to proceed with great caution in regard to whether the Project itself can be conceptualised in the same way. When Owen (2006b: 911) refers to the Project as a 'supranational *organisation*' and *'not an actor'*, he refers in the latter case to individual human capacities for agency. However, as to whether we can identify the Project as a *social* actor, it should be noted that in Sibeon's (2004: 119) definition, he does *not* refer to such phenomena as international research programmes alongside, 'organisations of the state, private or voluntary sections, including interest groups, political parties'. It is not clear as to whether we could make such a claim for the Project. The term *'supra-national organisation'* therefore is arguably more appropriate. *Micro-macro,* referring to '*relative* differences in size' and to *relational* conceptions of scale associated with *emergence* (Sibeon, 2004: 46) was applied to the Genome Project. We saw how it was argued that the Project can be viewed as *macro* 'Big Science' in which venture capitalists, researchers and politicians interact in a global sense. On a *micro* level, the Project is identified as being 'embedded' within what Bucchi (2004: 134) refers to as small-scale, 'heterogenous social networks'. In terms of *time-space* concepts, we 'applied' Sibeon's (2004: 166) concept of '*locale*' as a 'bridge' between *micro, mezo* and *macro* levels of interaction to the Genome Project. Here it was suggested that we focus upon the *internal* aspects of local sites (given that the sequencing of the genome was the culmination of over a decade of work involving 20 sequencing centre 'sites' in six countries). We also applied the concepts of *materials*, *material diffusion* and *networks* to the six techniques of human biotechnology (in general terms) identified by Lynn (2001) and discussed in depth in Chapter Four. We identified *discourse* pertaining to cloning, IVF, pre-natal screening, egg donation etc as *material*, existing temporally/spatially across a variety of *micro* sites and *macro* locales. We then identified how it would be possible to analyse *micro* (interactional) processes related to the application/handling of *material* within locales; and to study the human and social actors involved in *material diffusion* across time-space.

To reiterate, it is the contention here that the 'application' of meta-concepts to selected examples of human biotechnology and related biosocial areas, such as critiques of the bio-medical model (Powell and Owen, 2005); and attempts to 'build bridges' between the social and life sciences by those of the 'embodied' school such as Shilling (1993) and Newton (2003), has illustrated the considerable explanatory potential of the metatheoretical framework. The metatheoretical framework, having been 'modified' to include such terms as '*the oversocialised gaze'*, '*genetic fatalism*' and '*the biological variable*' is now better equipped to conceptualise areas of theoretical interest in human biotechnology than *oversocialised* or *under-socialised* frameworks. It is contended here that a framework which entails a flexible ontology, relying upon multi-factoral analysis rather than unitary, *reductionist, essentialist* and *reified* analysis is best equipped to point a 'way forward' beyond anti-foundational *relativism.* An approach which 'side-steps' the 'nature versus nurture' divide which still haunts mainstream social science, emphasising instead the *mutuality* between genes (nature) and environment (nurture) is best equipped to point a 'way forward' beyond Shilling's (1993) 'starting point' for a '*biological sociology*' (which appears to rely upon an elisionist, duality of biology/'the social'). These elements combined arguably make the framework one which can meaningfully contribute towards a 'return to' sociological theory and method as part of the *post-postmodern* 'school' associated with Sibeon (1996, 1999, 2004, 2007), Layder (1997, 2004, 2007), Archer (1988, 1995, 1998) , Mouzelis (1991, 1993, 1995, 2007) and Owen [2006a, 2006b, 2007a, 2007b]. Its methodological generalisations, as opposed to substantive generalisations, lack of *'biophobia'* (Freese et al, 2003), and 'realist' social ontology make it a 'sensitising device' with the potential for future theoretical and explanatory use best expressed in terms of large-scale synthesis.

References

Albrow, M. (1996) *The Global Age*. Cambridge: Polity Press.

Alexander, B. (2003) *Rapture: How Biotechnology Became the New Religion*. New York: Basic Books.

Alexander, J. (1992) 'General theory in the postpositivistic mode: the "epistemological dilemma" and the search for present reason", in S. Seidman and D.G. Wagner (eds), *Postmodernism and Social Theory*. Cambridge, MA, and Oxford: Blackwell.

Archer, M. (1982) 'Structuration versus morphogenesis: on combining structure and action', *British Journal of Sociology*, 33 (4): 445-83.

Archer, M. (1988*) Culture and Agency: The Place of Culture in Social Theory.* Cambridge: Cambridge University Press.

Archer, M. (1995) *Realist Social Theory: The Morphogenetic Approach,* Cambridge: Cambridge University Press.

Archer, M. (1996) 'Social integration and system integration: developing the distinction', *Sociology*, 30 (4): 679-99.

Archer, M. (1998) 'Social theory and the analysis of society', in T. May and M. Williams (eds), *Knowing the Social World.* Buckingham: Open University Press.

Archer, M. (2000) *Being Human: The Problem of Agency*. Cambridge: Cambridge University Press.

Armstrong, D. (1987) 'Theoretical terrains in biopsychosocial medicine', *Social Science and Medicine*, 25: 1213-18.

Association of British Insurers (1997) *Life Insurance and Genetics, A Policy Statement.*

Barkow, J. et al (1992) *The Adapted Mind.* Oxford: Oxford University Press.

Barnes, B. and Shapin, S. (eds) (1979) *Natural Order: Historical Studies of Scientific Culture*. London: Sage.

Barrett, L., Dunbar, R.I.M., and Lycett, J. (2002) *Human Evolutionary Psychology.* London: Belgrave.

Barthes, R. (1967) *Elements of Semiology*. London: Cape.

Baudrillard, J. (1983) *In the Shadow of the Silent Majorities or the End of the Social and Other Essays*. New York: Semiotext(e).

Bauman, Z. (1992) *Intimations of Postmodernity*. London: Routledge.

Beatty, J. (2000) 'Origins of the U.S. Human Genome Project: The Changing Relationship Between Genetics and National Security' in Sloan, P.R. (ed) (2000) *Controlling Our*

Destinies: Historical, Philosophical, Ethical, and Theological Perspectives on the Human Genome Project. Notre Dame: Notre Dame Press.

Beck, U. (1992) *The Risk Society*. London: Sage.

Beck, U. (1999) *World Risk Society*. Cambridge: Polity Press.

Beck, U. (2002) 'The silence of words and the political dynamics in the World Risk Society', *Logos,* 1(4):1-18.

Benjamin, J., et al (1996) 'Population and Familial Association Between the D4 Dopamine Receptor Gene and Measures of Novelty Seeking', in *Nature* 12: 81-84.

Benton, T. (1991) 'Biology and social science: why the return of the repressed should be given a (cautious) welcome'. *Sociology*, 25, 1: 1-29.

Benton, T. (1993) *Natural Relations: Ecology, Animal Rights and Social Justice.* London: Verso.

Benton, T. [1994] 'Biology and social theory in the environmental debate', in M.Redclift and T.Benton [eds.] [1994] *Social Theory and the Global Environment.* London: Routledge.

Benton, T. (1998) 'Realism and social science: some comments on Roy Bhaskar's 'The Possibility of Naturalism'', in Archer, M., Bhaskar, R., Collier, A., Lawton, T. and Norrie, A. (eds), *Critical Realism: Essential Readings*. London: Routledge.

Benton, T. (1999) 'Evolutionary Psychology: A new paradigm or just the same old reductionism?', *Advances in Human Ecology*, 1999.Benton, T. and Craib, I. (2001) *The Philosophy of Science*. London: Palgrave.

Benton, T. (2003) 'Ecology, health and society: a red-green perspective', in Williams, S.J., Birke, L. and Bendelow, G. (eds), *Debating Biology: Sociological Reflections on Health, Medicine and Society.* London: Routledge.

Bennington, J. and Harvey, J. (1994) 'Spheres or tiers? The significance of transnational local authority networks'. *Local Government Policy Making*, 20, 5: 21-30.

Bennington, J. and Taylor, M. (1993) 'Changes and challenges facing the UK welfare state in the EU of the 1990s'. *Policy* and *Politics*, 21, 2:121-34.

Betts, K. (1986) 'The conditions of action, power and the problem of interests', *Sociological Review*, 34(1): 39-64.

Berger, P. and Luckmann, T. (1971) *The Social Construction of Reality.* Harmondsworth: Penguin. .

Best, S. and Kellner, D. (1991) *Postmodern Theory: Critical Interrogations.* London: Macmillan.

Bhaskar, R. (1989a) *Reclaiming Reality: A Critical Introduction to Contemporary Philosophy.* London: Verso.

Bhaskar, R. (1989b) *The Possibility of Naturalism*. London: Harvester Wheatsheaf.

Biggs, S. (1999) *The Mature Imagination*. OUP: Milton Keynes.

Biggs, S. and Powell, J.L. (1999) 'Surveillance and Elder Abuse: The Rationalities and Technologies of Community Care', *Journal of Contemporary Health*, 4(1), 43-49.

Bijker, W. (1995) *Of Bicycles, Bakelites and Bulbs*. New York: Routledge.

Birke, L. (2003) 'Shaping biology: Feminism and the idea of 'the biological'', in Williams, S.J., Birke, L. and Bendelow, G. (eds), *Debating Biology: Sociological Reflections on Health, Medicine and Society*. London: Routledge.

Blackledge, D. and Hunt, B. (1993) *Sociological Interpretations of Education.* London: Routledge.

Bloor, D. (1976) *Knowledge and Social Imagery*. London: Routledge and Kegan Paul, 2nd edn, Chicago: Chicago University Press, 1991.

Bodmer, W.F. and McKie, R. (1994) *The Book of Man*. London: Little Brown.

Bogaert, A.F. and Fisher, W.A. (1995) 'Predictors of University Men's Number of Sexual Partners'. *The Journal of Sex Research,* 32, 119-30.

Bojorquez, J. (1994) 'In his image'. *Sacramento Bee*, p.18.

Boseley, S. (2005) 'Drive launched to find egg and sperm donors', *Guardian*, Wed 26 January.

Bourdieu, P. [1977] *Outline of a Theory of Practice*. Cambridge: Cambridge University Press.

Bourdieu, P. [1984] *Distinction: A Social Critique of the Judgement of Taste*. Cambridge, MA: Harvard University Press.

Bourdieu, P. [1990] *In Other Words: Essays Towards a Reflexive Sociology*. Cambridge: Polity Press.

Bourdieu, P. [1998] *Practical Reason: On the Theory of Action*. Cambridge: Polity Press.

Brennan, Z. and Syal, R. (1997 August 17) 'Sperm donor payments to end'. *Sunday Times* (London), p.7.

Bretherton, C. [1996] 'Introduction: global politics in the 1990s'' in C. Bretherton and G.Ponton [eds.] *Global Politics: An Introduction*. Oxford: Blackwell.

Bridenthall, R., Grossman, A., and Kaplan, M. (eds) (1984) *When Biology Became Destiny: Women in Weimar and Nazi Germany*. New York: Monthly Review Press.

Brunner, H.G., et al (1993a) 'X-linked Borderline Mental Retardation with Prominent Behavioural Disturbance: Phenotype, Genetic Localization and Evidence for Disturbed Monoamine Metabolism'. *American Journal of Human Genetics* 52: 1032.

Brunner, H.G., et al (1993b) 'Abnormal Behaviour Associated with a Point Mutation in the Structural Gene for Monoamine Oxidose A', *Science* 262: 578-80.

Brittan, S. (1998) 'Essays, moral, political and economic'. *Hume Papers on Public Policy,* vol. 6, no. 4. Edinburgh: Edinburgh University Press.

Bromley, D.W. and Segerson, K. (1992) (eds) *The Social Response to Environmental Risk: Policy Formation in an Age of Uncertainty*. London: Springer.

Brodwin, P.E. (2001) (ed) *Biotechnology and Culture: Bodies, Anxieties, Ethics*. Indiana: Indiana University Press.

Brown, K. (2000) 'The Human Genome Business Today'. *Scientific American*, 283(1): 7.

Brown, R.S. and Marshall, K. (1993) (eds) *Advances in Genetic Information: A Guide for State Policy Makers*. Lexington, KY: Council of State Governments.

Bucchi, M. (2004) *Science in Society: An Introduction to Social Studies of Science*. London and New York: Routledge.

Buctuanon, E.M. (2001) 'Globalisation of biotechnology: the agglomeration of dispersed knowledge and information and its implications for the political economy of technology in developing countries', *New Genetics and Society*, 20, (1).

Bulmer, M. (1985) 'The rejuvenation of community studies? Neighbours, networks and policy', *The Sociological Review*, 33(3): 430-48.

Bulmer, M. (1986) *Neighbours: The Works of Philip Abrams*. Cambridge: Cambridge University Press.

Bury, M.R. [1995] 'The body in question', *Medical Sociology News,* 21 [1]:36-48.

Bury, M.R. (1997) *Health and Illness in a Changing Society*. London: Routledge.

Buss, D.M. (1999) *Evolutionary Psychology*. Boston: Allyn and Bacon.

Caldwell, J.H. (1984) 'Babies by scientific selection', *Scientific American*, 150, 124-25.

Calhoun, C., et al (2002) (eds) *Contemporary Sociological Theory*. Oxford: Blackwell.

Callon, M. (1986) 'Some elements of a sociology of translation: domestication of the scallops and the fisherman of St. Brieuc Bay', in J. Law (ed), *Power Action and Belief: A New Sociology of Knowledge?* London: Routledge.

Callon, M and Latour, B. (1981) 'Unscrewing the big Leviathan: how actors macro-structure reality and how sociologists help them to do so', in K. Knorr-Cetina and A.V. Cicourel (eds) *Advances in Social Theory and Methodology: Towards an Integration of Micro-and Macro-Sociologies*. London: Routledge.

Cannon, W.B. (1914) 'The interrelations of emotion as suggested by recent physiological researchers', *American Journal of Psychology*, 25: 256-282.

Cantor, C. (1992) 'The challenges to technology and information' in D.J. Kevles and L. Hood (eds) *The Code of Codes*. Cambridge, MA: Harvard University Press.

Cantor, J.M., Blanchard, R., Paterson, A.D., and Bogaert, A.F. (2002) How many gay men owe their sexual orientation to fraternal birth order? *Archives of Sexual Behaviour*, 31: 63-71.

Caplan, A.L. (2000) 'What's Morally Wrong with Eugenics?' in Sloan, P.R. (ed) *Controlling our Destinies: Historical, Philosophical, Ethical, and Theological Perspectives on the Human Genome Project.* Notre Dame: Notre Dame Press.

Cazillis, M. (2001) 'Facing omnipresent ethical problems', EUROPA, European Commission Research Information Centre, update 1/12/2001.

Clegg, S. (1989) *Frameworks of Power.* London: Sage.

Churchland, S. and Grush, R. (1999) 'Computation and the Brain', in R.A. Wilson and F.C. Kiel (eds) *The MIT Encyclopedia of the Cognitive Sciences.* Cambridge, MA: MIT Press.

Coburn, D. (2000a) 'Income inequality, social cohesion and the health status of populations: the role of neoliberalism', *Social Science and Medicine,* 51: 135-146.

Coburn, D. (2000b) 'A brief response', *Social Science and Medicine,* 51: 1009-1010.

Cohen, C. [1996] *New Ways of Making Babies: The Case of Egg Donation.* Bloomington: Indiana University Press.

Cohen, I. [1987] 'Structuration theory and social praxis', in A.Giddens and J.Turner [eds.] *Social Theory Today.* Cambridge: Polity Press.

Cohen, I. (1989) *Structuration Theory: Anthony Giddens and the Constitution of Social Life.* London: Macmillan.

Collins, H. and Kusch, M. (1998) *The Shape of Actions: What Humans and Machines Can Do*. Cambridge, MA: MIT Press.

Collins, H.M., and Pinch, T. (1993) *The Golem: What Everyone Should Know About Technology*. Cambridge: Cambridge University Press.

Cosmides, L. and Tooby, J. (1992) 'Cognitive adaptations for social exchange', in Barkow, J., Cosmides, L. and Tooby, J. *The Adapted Mind: Evolutionary Psychology and the Generation of Culture*. New York: Oxford University Press.

Cosmides, L. and Tooby, J. (1997) *Evolutionary Psychology: A Primer*. Available at the web site http://www.psych.ucsb.edu/research/cep/primer.html.

Crawford, C.B. and Anderson, J.L. (1989) 'Sociobiology: an environmentalist discipline?', *American Psychologist*, 44(12): 1449-59.

Curry, O. (2003) 'Evolutionary psychology: "fashionable ideology" or "new foundation"?', *Human Nature Review*, 3: 81-92 (28 January).

Daly, M. and Wilson, M. (1988) *Homicide*. New York: Aldine de Gruyter.

Daly, M. and Wilson, M. (1998) *The Truth About Cinderella: A Darwinian View of Parental Love*. London: Weidenfeld and Nicolson.

Daly, M. and Wilson, M. (2001) 'An assessment of some proposed exceptions to the phenomenon of nepotistic discrimination against stepchildren', *Annales Zoologici Fennici,* 38: 287-296.

Danaher, G., et al (2000) *Understanding Foucault*. London: Sage.

David, M. (2002) 'The sociological critique of evolutionary psychology: beyond mass modularity', *New Genetics and Society*, 21, 3.

David, M. and Kirkhope, J. (2005) 'Cloning/stem cells and the Meaning of Life', *Current Sociology*, 53(2), 367-381.

Davies, I. [1993] 'Cultural theory in Britain: narrative and episteme', *Theory, Culture and Society,* 10 [3]:115-54.

Dawkins, R. (1976) *The Selfish Gene*. Oxford: Oxford University Press.

Dawkins, R. (1986) *The Blind Watchmaker*. New York: W.W. Norton and Co.

Dawkins, R. (1992) 'Progress' in E. Fox Keller and E. Lloyd (eds) *Keywords in Evolutionary Psychology*. Cambridge, Mass: Harvard University Press.

Degener, T. (1990) 'Female Self-determination Between Feminist Claims and "Voluntary" Eugenics, between "Rights" and Ethics, Issues in Reproductive and Genetic Engineering', *Journal of International Feminist Analysis*, 3(2) 87-99.

Delanty, G. (1999) *Social Theory in a Changing World: Conceptions of Modernity*. Cambridge: Polity Press.

Delanty, G. (2002) 'Constructivism, sociology and the new genetics', *New Genetics and Society,* 21, 3: 279-289.

Dennis, C. and Gallagher, R. (eds) (2001) *The Human Genome*. Hampshire and New York: Palgrave.

Derbyshire, D. and Highfield, R. (2003) 'Women better than men at instant maths', the *Daily Telegraph*, 11th September.

Derrida, J. (1982) 'Différence', in *Margins of Philosophy*, trans. A. Bass, Hemel Hempstead: Harvester Wheatsheaf.

Dickens, P. [1996] *Reconstructing Nature: Alienation, Emancipation and the Division of Labour*. London: Routledge.

Dickens, P. (2000) *Social Darwinism*. Buckingham: Open University Press.

Dietz, T.R., Frey, R.S. and Rosa, E. [1997] 'Risk, technology and society', in R.E.Dunlap and W.Michelson [eds.] *Handbook of Environmental Sociology*. Westport, CT: Greenwood Press.

Dominelli, L. (1997) *Sociology for Social Work*. London: Macmillan.

Dreger, A.D. (2000) 'Metaphors of Morality in the Human Genome Project', in Sloan, P.R. (ed) *Controlling Our Destinies: Historical, Philosophical, Ethical, and Theological Perspectives on the Human Genome Project*. Notre Dame: Notre Dame Press.

Dunbar, R.I.M. (1996) *Gossip, Grooming and the Evolution of Language*. London: Faber and Faber.

Dunbar, R.I.M. (2003a) 'Why are apes so smart?' in P. Keppeler and M. Pereira (eds) *Primate Life Histories and Socioecology*. Chicago: Chicago University Press.

Dunbar, R.I.M. (2003b) 'The social brain: mind, language and society in evolutionary perspective' in *Annual Review of Anthropology*, 32: 163-181.

Dunbar, R.I.M. (2004) *The Human Story*. London: Faber and Faber.

Dunbar, R.I.M. (2006) 'Taking Social Intelligence Seriously', in *Human Ability: Genetic and Environmental Influences,* proceedings of a conference organised by The Galton Institute, London, 2005. Edited by Robert Peel and Marzin Zekin. Published by The Galton Institute.

Dunbar, R.I.M, Knight, C. and Power, C. (1999) *The Evolution of Culture*. Edinburgh: Edinburgh University Press.

Dupre, J. (2001) *Human Nature and the Limits of Science*. Oxford: Oxford University Press.

Durant, J., Baur, M.W., and Gaskell, G. (1998) *Biotechnology in the Public Sphere: A European Sourcebook*. London: NMS Trading.

Durkheim, E. (1982) *The Rules of Sociological Method*. London: Macmillan.

Duster, T. (1990) *Backdoor to Eugenics*. New York: Routledge.

Dye, T. (1998) *Understanding Public Policy*. Upper Saddle River, NJ: Prentice Hall.

Enard, W., Przeworski, M., Fisher, S.E., Lai, C.S.L., Wiebe, V., Kitano, T., Monaco, A.P. and Paabo, S. (2002) 'Molecular evolution of FOXP2, a gene involved in speech and language', *Nature,* 418: 869-72.

Elias, N. (1978) *What is Sociology?* London: Hutchison.

Elliot, P. and Mandell, N. (1995) 'Feminist theories', in N. Mandell (ed) *In Answering Back: Girls, Boys and Feminism in Schools*. Scarborough, ON: Prentice Hall Canada Inc.

Elstad, J.I. (1998) 'The psycho-social perspective on social inequalities in health', *Sociology of Health and Illness*, 20(5): 598-618.

Eltzinga, A. and Jamison, A. (1995) 'Changing Policy Agendas in Science and Technology', in S. Jasonoff et al (eds) *Handbook of Science and Technology Studies*. Thousand Oaks: Sage.

Etzkowitz, H. and Webster, A. (1995) 'Science as Intellectual Property', in S. Jasonoff et al (eds) *Handbook of Science and Technology Studies*. Thousand Oaks: Sage.

Ericson, R.V., and Doyle, A. (eds) (2003) *Risk and Morality*. Toronto: University of Toronto Press.

Ettorre, E. [2000] 'Reproductive genetics, gender and the body: 'Please doctor, may I have a normal baby?', *Sociology,* 34 [3]:403-20.

European Parliament (1990) *Ethical and Legal Problems of Genetic Engineering and Human Artificial Insemination*. Strasbourg.

European Parliament (1997, March 11). 'Resolution on cloning', *Proceedings of the European Parliament*. Strasbourg.

Ewing, C.M. (1998) 'Tailored Genes: IVF, Genetic Engineering and Eugenics, Reproductive and Genetic Engineering', *Journal of International Feminist Analysis*, 3(2): 119-23.

Ezzell, C. (1987) 'First ever animal patent issues in United States', *Nature*, 332, 21.

Farganis, S. (1994) *Situating Feminism: From Thought to Action*. Thousand Oaks, CA, and London: Sage.

Farganis, S. [1994] *Situating Feminism: From Thought to Action*. Thousand Oaks, CA, and London: Sage.

Featherstone, M. (1988) 'In pursuit of the postmodern', *Theory, Culture and Society,* 5: 195-215.

Featherstone, M. and Wernick, A. (1995) *Images of Ageing*. London: Routledge.

Feingold, W. (1976) *Artificial Insemination*. New York: Academic Press.

Fleising, U. (2001) 'Genetic essentialism, *mana*, and the meaning of DNA', *New Genetics and Society*, 20(1): 43-57.

Fodor, J. (1983) *The Modularity of Mind*. Cambridge, MA: MIT Press.

Fodor, J. (2000) *In Critical Condition*. Cambridge, MA: MIT Press.

Fodor, J. (2001) *The Mind Doesn't Work That Way*. Cambridge, MA: MIT Press.

Foucault, M. (1970) *The Order of Things: An Archaeology of the Human Sciences*. London: Tavistock.

Foucault, M. (1972) *The Archaeology of Knowledge*. New York: Pantheon Books.

Foucault, M. (1980a) *Power/Knowledge*. New York: Pantheon Books.

Foucault, M. (1980b) *The History of Sexuality*. New York: Vintage Books.

Foucault, M. (1982) 'The subject and power', in H.L. Dreyfus and P. Rabinow (eds), *Michael Foucault: Beyond Structuralism and Hermeneutics, with an Afterword by Michel Foucault*. Brighton: Harvester Wheatsheaf.

Foucault, M. (1991) 'Governmentality', in G. Burchell (ed), *The Foucault Effect: Studies in Governmentality*. Hemel Hempstead: Harvester Wheatsheaf.

Fox, C.J. and Miller, H.T. (1995) *Postmodern Public Administration*. London: Sage.

Frank, A.W. (1990) 'Bringing Bodies Back In: A Decade Review', *Theory, Culture and Society*, 7, 1: 131-162.

Freese, J., Li, J.C.A. and Wade, L.D. (2003) 'The potential relevances of biology to social inquiry', *Annual Review of Sociology*, 29: 233-256.

Freund, P.E.S. (1988) 'Bringing society into the body', *Theory and Society*, 17: 839-864.

Freund, P.E.S. (1990) 'The expressive body: a common ground for the sociology of emotions and health and illness', *Sociology of Health and Illness*, 12, 4: 452-477.

Freund, P.E.S. (1998) 'Social performances and their discontents: the biopsychosocial aspects of dramaturgical stress', in Bendelow, G. and Williams, S.J. (eds) *Emotions in Social Life: Critical Themes, Contemporary Issues*. London: Routledge.

Freund, P.E.S. (2001) 'Bodies, disability and spaces: the social model of disability and disabling spatial organisations', *Disability and Society*, 16, 5.

Friedrichs, R.W. (1972) *A Sociology of Sociology*. New York: Free Press.

Gagnon, J.H. and Simon, W. (1973) *Sexual Conduct*. London: Hutchinson.

Galton Institute [1999] 'Notes of the quarter', *The Galton Institute Newsletter*, [2]:8.

Gellner, E. (1993) *Postmodernism, Reason and Religion*. London and New York: Routledge.

Gibbs, J.P. (1989) *Control: Sociology's Central Notion*. Urbana, IL: University of Illinois Press.

Gibbons, M., Limoges, C., Nowotny, H. et al (1994) *The New Production of Knowledge: The Dynamics of Science and Research in Contemporary Societies*. London: Sage.

Giddens, A. (1979) *Central Problems in Social Theory: Action, Structure and Contradiction in Social Analysis*. Berkeley, CA: University of California Press.

Giddens, A. (1981) 'Agency, institution and time-space analysis', in K.Knorr-Cetina and A.V. Cicourel (eds) *Advances in Social Theory and Methodology: Towards an Integration of Micro-and Macro-Sociologies*. London: Routledge.

Giddens, A. (1982) *Profiles and Critiques in Social Theory*. London: Macmillan.

Giddens, A. (1984) *The Constitution of Society*. Cambridge: Polity Press.

Giddens, A. (1987) *Social Theory and Modern Sociology*. Cambridge: Polity Press.

Giddens, A. (1989) ‘A reply to my critics’ in D. Held and J. Thompson (eds) *Social Theory and Modern Societies: Anthony Giddens and his Critics*. Cambridge: Cambridge University Press.

Giddens, A. (1991a) ‘Structuration theory: past, present and future’, in C. Bryant and D. Jary (eds) *Giddens's Theory of Structuration: A Critical Appreciation*. London: Routledge.

Giddens, A. (1991b) *Modernity and Self-Identity*. Cambridge: Polity Press.

Giddens, A. (1993) *Sociology*. Cambridge: Polity Press.

Gigerenzer, G., Todd, P., and the ABC Research Group (1999) *Simple Heuristics that Make US Smart*. Milton Keynes: Open University Press.

Goffman, E. (1983) ‘The interaction order’, *American Sociological Review*, 48(3): 1-17.

Gottweis, H. (2005) ‘Governing genomics in the 21st century: between risk and uncertainty’, *New Genetics and Society*, 24, 2: 175-194.

Gurnah, A. and Scott, A. (1992) *The Uncertain Science: Criticisms of Sociological Formalism*. London: Routledge.

Habermas, J. [1986] *The Theory of Communicative Action. Vol. 1: Reason and the Rationalization of Society*. Cambridge: Polity Press.

Habermas, J. (1987) *The Theory of Communicative Action. Vol. II: The Critique of Functionalist Reason*. Cambridge: Polity Press.

Habermas, J. [1989] *The Structural Transformation of the Public Sphere*. Cambridge, MA:MIT Press.

Hall, S.S. (2003) *Merchants of Immortality: Chasing the Dream of Human Life Extension*. New York: Houghton Miffin.

Hamer, D. and Copeland, P. (1999) *Living With Our Genes: Why They Matter More Than You Think*. London: Macmillan.

Hamerton, J.L., Evans, J.A., and Stranc, L. (1993) ‘Prenatal diagnosis in Canada’, in *Research Volumes of the Royal Commission on New Reproductive Technologies*. Ottawa: H.M.S.O.

Hanmer, J. (1993) ‘Reproductive Technology: The Future for Women’, in J. Rothchild (ed), *Machina Ex Dea: Feminist Perspectives on Technology*. New York: Pergamon.

Hannigan, J.A. (1995) *Environmental Sociology: A Social Constructionist Perspective*. London: Routledge.

Haraway, D. (1997) *Modest_Witness @ Second Millenium. Female Man Meets Oncomouse TM*. New York: Routledge.

Harris, J.R. (1998) *The Nurture Assumption*. London: Bloomsbury.

Harvey, D. (1989) *The Condition of Postmodernity*. Oxford: Blackwell.

Harre, R. (1981) ‘Philosophical aspects of the macro-micro problem’ in K.C. Knorr-Cetina and A.V. Cicourel (eds) *Advances in Social Theory and Methodology: Towards an Integration of Micro-And Macro-Sociologies*. London: Routledge.

Hay, C. (2002) *Political Analysis: A Critical Introduction*. Basingstoke: Palgrave.

Hayek, F.A. (1960) *The Constitution of Liberty*. London: Routledge and Kegan Paul.

Hindess, B. (1986a) ‘Actors and social relations’, in M.L. Wardell and S.P. Turner (eds), *Sociological Theory in Transition*. London: Allen and Unwin.

Hindess, B. (1986b) ‘Interests in political analysis’, in J. Law (ed), *Power Action and Belief: A New Sociology of Knowledge?* London: Routledge.

Hindess, B. (1988) *Choice. Rationality and Social Theory*. London: Unwin Hyman.

Hochschild, A.R. (1983) *The Managed Heart: Commercialisation and Human Feeling.* Berkeley: University of California Press.
Hochschild, A.R. (1990) *The Second Shift: Working Parents and the Revolution at Home.* London: Piatkus.
Holmwood, J. (1996) *Founding Sociology? Talcott Parsons and the Idea of General Theory.* London and New York: Longman.
Holzner, B. (1978) 'The construction of social actors: an essay on social identities', in T. Luckmann (ed), *Phenomenology and Sociology.* Harmondsworth: Penguin.
Irwin, A. (2001) *Sociology and the Environment.* Cambridge; Polity Press.
Irwin, A. (2003) (ed) and Lynne, B. (ed) *Misunderstanding Science?: The Public Reconstruction of Science and Technology.* Cambridge: Cambridge University Press.
James, S. (1997) *British Government: A Reader in Policy Making.* London: Routledge.
Jameson, F. (1991) Postmodernism or the Cultural Logic of Late Capitalism. London: Verso.
Jones, M. (1996) 'Posthuman agency: between theoretical traditions', *Sociological Theory,* 14(3): 290-309.
Jones, M. and Salter, B. (2004) 'The governance of human genetics: policy discourse and constructions of public trust', *New Genetics and Society*, 22(1), 4: 21-41.
Kandel, E., Schwartz, J. and Jessel, T. (1995) *Principles of Neural Science.* New York: Elsevier.
Karmiloff-Smith, A. (1992) Beyond Modularity: A Developmental Perspective on Cognitive Science. Cambridge, MA: MIT Press.
Karmiloff-Smith, A. (2000) 'Why babies' brains are not Swiss Army knives', in H. Rose and S. Rose (eds) *Alas Poor Darwin: Arguments against Evolutionary Psychology.* London: Jonathan Cape.
Katz, S. (1997) *Disciplining Old Age*. Charlottesville: University of Virginia.
Kaupen-Haas, H. (1988) 'Experimental Obstetrics and National Socialism: The Conceptual Basis of Reproductive Technology Today', *Reproductive and Genetic Engineering: Journal of International Feminist Analysis*, 1(2): 127-32.
Keller, E.F. (1995) *Refiguring Life: Metaphors of Twentieth-Century Biology.* New York: Columbia Press.
Kelsoe, J., Ginns, E., Egeland, J., et al (1989) 'Reevaluation of the linkage relationship between chromosome 11p loci and the gene for Bipolar Affective Disorder in the Old Order Amish', *Nature,* 342: 238-43.
Kerr, A. (2003) 'Rights and responsibilities in the new genetics era', *Critical Social Policy*, 23(2): 208-226.
Kevles, D.J. (1985) *In the Name of Eugenics*. New York: Knopf.
Kevles, D.J. and Hood, L. (1992) 'Reflections', in D.J. Kevles and L. Hood (eds), *The Code of Codes*. Cambridge, M.A.: Harvard University Press.
King, T.J., and Briggs, R. (1956) 'Serial transplantation of embryonic nuclei', *Cold Spring Harbor Symposium on Quantitative Biology*, 21: 271-90.
Kitcher, P. (1996) *The Lives to Come*. New York: Simon and Schuster.
Kolata, G. (1997) *Clone*. London: Alton Lane.
Kollek, R. (1990) 'The Limits of Experimental Knowledge: A Feminist Perspective on the Ecological Risks of Genetic Engineering', *Issues in Reproductive and Genetic Engineering. Journal of International Feminist Analysis*, 3(2): 125-35.
Kooiman, J. (2003) *Governing as Governance*. London: Sage.

Knorr-Cetina, K.D. and Cicourel, A.V. (1981) (eds) *Advances in Social Theory and Methodology: Towards an Integration of Micro-And Macro-Sociologies.* London: Routledge

Knorr-Cetina, K. (1981) *The Manufacture of Knowledge: An Essay on the Constuctivist and Contextual Nature of Science.* Oxford: Pergamon.

Koshland, D.E. (1988) 'The future of biological research: What is possible and what is ethical?' *Science*, 3, 11-15.

Koski, C.A. (2005) 'The Human Genome Project: an examination of its challenge to the technological imperative', *New Genetics and Society*, 24(3): 267-281.

Kumar, K. (1995) *From Post-Industrial to Post-Modern Society: New Theories of The Contemporary World.* Oxford: Blackwell

Kupferman, I. (1992) 'Genetic Determinants of Behaviour', in Kendel, E., Schwartz, J. and Jessel, T. (1995) *Principles of Neural Science.* New York: Elsevier.

Kuhn, T.S. (1962) *The Structure of Scientific Revolutions.* Chicago: Chicago University Press (2nd edn. 1969).

Kurzban, R. (2002) 'Alas poor evolutionary psychology: unfairly accused, unjustly condemned', *Human Nature Review*, 2: 99-109. (Available at http://human-nature.com/nibbs/02/apd.htm).

Lacan, J. (1977) *Ecrits: A Selection.* London: Tavistock.

Laclau, E. and Mouffe, C. (1985) *Hegemony and Socialist Strategy.* London: Verso.

Laclau, E. and Mouffe, C. (1987) 'Post-Marxism without apologies', *New Left Review,* 166: 79-106.

Lai, C.S., Fisher, S.E. et al (2001) 'A forkhead-domain gene is mutated in a severe speech and language disorder', *Nature*, 413: 519-23.

Latour, B. (1986) 'The powers of association', in J. Law (ed), *Power Action and Belief: A New Sociology of Knowledge?* London: Routledge.

Latour, B. (1987) *Science in Action.* Cambridge, MA: Harvard University Press.

Latour, B. (1988) '*The Pasteurization of France.* Cambridge, MA: Harvard University Press.

Latour, B. (1993) *We Have Never Been Modern.* London: Harvester Wheatsheaf.

Latour, B. and Woolgar, S. (1979) *Laboratory Life: The Social Construction of Scientific Facts.* Princeton: Princeton University Press.

Lau, S. (1995) 'Report from Hong Kong', *Cambridge Quarterly of Healthcare Ethics*, 4: 364-66.

Law, J. (1986a) 'On power and its tactics: a view from the sociology of science', *Sociological Review,* 34(1): 1-38.

Law J. (1986b) *Power Action and Belief: A New Sociology of Knowledge?* London: Routledge

Law, J. (ed) (1991a) *A Sociology of Monsters: Essays on Power, Technology and Domination.* London: Routledge.

Law, J. (1991b) 'Power, discretion and strategy', in J. Law (ed) ibid.

Law, J. (1994) *Organising Modernity.* Oxford: Blackwell.

Law, J. (ed) (1999) *Actor Network Theory and After.* London: Blackwell.

Layder, D. (1984) 'Sources and levels of commitment in actors' careers', *Work and Occupations, 1*11(2): 198-216.

Layder, D. (1994) *Understanding Social Theory.* London: Sage.

Layder, D. (1997) *Modern Social Theory: Key Debates and New Directions.* London: UCL Press.

Layder, D. (1998a) *Sociological Practice: Linking Theory and Social Research.* London: Sage.

Layder, D. (1998b) 'The reality of social domains: implications for theory and method', in T. May and M. Williams (eds), *Knowing the Social World.* Buckingham: Open University Press.

Layder, D. (2004) *Emotion in Social Life: The Lost Heart of Society.* London: Sage.

Layder, D. [2007] 'Self-identity and personhood in social analysis: the inadequacies of postmodernism and social constructionism', in J.L.Powell and T.Owen [eds.] *Reconstructing Postmodernism: Critical Debates.* New York: Nova Science Publishers.

Layton, ET. (1988) 'Science as a Form of Action. The Role of the Engineering Sciences', in *Technology and Culture*, 28: 594-607.

Lemert, C. (1993) 'After modernity', in C. Lemert (ed) *Social Theory: The Multicultural and Classical Readings.* San Francisco: Westview Press.

Lenoir, T. and Hays, M. (2000) 'The Manhattan Project for Biomedicine', in P.R. Sloan (ed) *Controlling Our Destinies: Historical, Philosophical, Ethical and Theological Perspectives on the Human Genome Project.* Notre Dame: Notre Dame Press.

Leuzinger, M. and Rambert, B. [1988] 'I can feel it- my baby is healthy: women's experiences with pre-natal diagnosis in Switzerland', *Reproductive and Genetic Engineering,* 1 [3]: 239-49.

Levi-Strauss, C. (1963) *Structural Anthropology.* New York: Basic Books.

Levi-Strauss, C. (1974) *The Savage Mind.* London: Weidenfeld and Nicholson.

Lewontin, R. (1992) 'The Dream of the Human Genome', *New York Review of Books,* 39: 31-40.

Lewontin, R. (2000) *It Ain't Necessarily So.* New York: Granta Books.

Lippman, A. (1991) 'Prenatal genetic testing and screening: constructing needs and reinforcing inequities', *American Journal of Law and Medicine,* 17, 15-50.

Lockwood, D. (1964) 'Social integration and system integration', in G.K. Zollschan and W.Hirsch (eds), *Explorations in Social Change.* London: Routledge.

Longino, C.F. and Powell, J.L. (2004) 'Embodiment and the Study of Ageing' in V. Berdayes (ed) *The Body in Human Inquiry: Interdisciplinary Explorations of Embodiment.* New York: Hampton Press.

Luhmann, N. (1982) *The Differentiation of Society.* New York: Columbia University Press.

Lupton, D. (1999) *Risk.* London: Routledge.

Lyotard, J.F. (1973) *La Fenomenologia.* Buenos Aires: EUDEBA.

Lyotard, J.F. (1986) *The Postmodern Condition: A Report on Knowledge,* trans, G. Bennington and B. Massumi. Manchester: Manchester University Press.

Lyotard, J.F. (1986-1987) 'Rules and paradoxes and svelte paradox', *Cultural Critique,* 5: 209-19.

Lynn, R. (2001) *Eugenics: A Reassessment.* Westport, Connecticut: Praeger.

Lyon, D. (1994) *Postmodernity.* Milton Keynes: Open University Press.

McCormick, R.A. [1994] 'Blastomere separation: some concerns', *Hastings Center Report* [24]: 14-16.

MacKenzie, D. [1978] 'Statistical theory and social interests: a case study', *Social Studies of Science,* [8]: 35-83.

MacKenzie, D. [1996] 'How do we know the properties of artefacts? Applying the sociology of knowledge to technology', in R.Fox [ed.] *Technological Change: Methods and Themes in the History of Technology.* Reading: Harwood.

MacKenzie, D. and Wajcman, J. [eds.] [1999] *The Social Shaping of Technology.* Buckingham: Open University Press.

McLennan, C. (1995) 'After postmodernism: back to sociological theory?' *Sociology,* 29(1): 117:32.

McNay, L. (1992) *Foucault and Feminism: Power, Gender and the Self.* Cambridge: Polity Press.

Mannheim, K. (1925) 'The Problem of a Sociology of Knowledge from a dynamic standpoint', in H. Nowotny and K. Taschwer (eds) *The Sociology of the Sciences*, vol. 1: 3-14.

Marsh, D. (ed) (1998) *Comparing Policy Networks*. Buckingham: Open University Press.

Marsh, D. and Smith, M. (2000) 'Understanding policy networks: towards a dialectical approach'. *Political Studies*, 48(1): 4-21.

Marsh, D. and Stoker, G. (eds) (1995) *Theory and Methods in Political Science*. London: Macmillan.

Martin, P. and Frost, R. (2003) 'Regulating the commercial development of genetic testing in the UK: problems, possibilities and policy', *Critical Social Policy*, 23(2): 187-207.

May, T. (1996) *Situating Social theory*. Milton Keynes: Open University Press.

Mead, G.H. (1967) *Mind, Self and Society*. Chicago: University of Chicago Press.

Merriden, T. [2001] *Irresistable Forces: The Legacy of Napster and the Growth of the Underground Internet.* Oxford: Capstone.

Merton, R.K. (1938) *Science, Technology and Society in Seventeenth-Century England.* Bruges: St Catherine Press (fourth edn, with a new introduction. New York, Howard Fertig, 2001).

Merton, R.K. (1942) 'The Normative Structure of Science', repr. in *The Sociology of Science*, 1973.

Merton, R.K. (1952) 'The Neglect of the Society of Science', repr. in *The Sociology of Science*, 1973.

Merton, R.K. (1968) *Social Theory and Social Structure* (original edn. 1949), New York: Free Press.

Midgley, M. (2000) *Science and Poetry.* London: Routledge.

Mill, J.S. (1859) *On Liberty*. London: Macmillan.

Miller, S. [1994] 'Wrinkles, ripples and fireballs: cosmology on the front page', *Public Understanding of Science,* [3]: 445-453.

Mouzelis, N. (1989) 'Restructuring structuration theory'. *The Sociological Review*, 37(4): 613-35.

Mouzelis, N. (1991) *Back to Sociological Theory: The Construction of Social Order*. London: Macmillan.

Mouzelis, N. (1993a) 'The poverty of sociological theory', *Sociology,* 27(4): 675-95.

Mouzelis, N. (1993b) 'Comparing the Durkheimian and Marxist traditions', *The Sociological Review*, 41(3): 572-82.

Mouzelis, N. (1994) 'In defence of "grand" historical sociology', *British Journal of Sociology*, 45(1): 31-6.

Mouzelis, N. (1995) *Sociological Theory: What Went Wrong? Diagnosis and Remedies.* London: Routledge.

Mouzelis, N. (1996) 'After postmodernism: a reply to Gregor McLennan', *Sociology,*_30(1): 131-35.

Mouzelis, N. (1997) 'Social and system integration: Lockwood, Habermas, Giddens', *Sociology,* 31(1): 111-19.

Mouzelis, N. (2000) 'The subjectivist-objectivist divide: against transcendence', *Sociology*, 34(4): 741-62.

Mouzelis, N. [2007] 'Cognitive relativism: between positivistic and relativistic thinking in the social sciences', in J.L.Powell and T.Owen [eds.] *Reconstructing Postmodernism: Critica Debates.* New York: Nova Science Publishers.

Munch, R. and Smelser, N. (1987) 'Relating the micro and macro', in J. Alexander, B. Giesen, R. Munch and N. Smelser (eds), *The Micro-Macro Link.* Berkeley, CA: University of California Press.

Nelkin, D. (1994) 'Promotional metaphors: the gene in popular discourse', *Nature Review: Genetics*, 2: 555-559.

Nelkin, D. (2001) 'Anything for an edge: breeding a race of champions by germline', *The Galton Institute Newsletter,* 43: 5-6.

Nesse, R. (1990) 'Evolutionary Explanations of Emotions', *Human Nature,* 1 (3): 261-289.

Nesse, R.M. and Williams, G.C. (1994*) Evolution and Healing: The New Science of Darwinian Medicine*. London: Pheonix.

Newman, F. and Brody, E.B. (1988) 'Psychosocial and ethical concerns in new reproductive technologies', Editorial of *Journal of Psychosomatic Obstetric Gynaecology,* 9: 155-8.

Newton, T. (2003) 'Truly embodied sociology: marrying the social and the biological?', *Sociological Review,* 51(1): 20-41.

Nicholson, L. and Seidman, S. (1995) 'Introduction', in L. Nicholson and S. Seidman (eds), *Social Postmodernism: Beyond Identity Politics.* Cambridge: Cambridge University Press.

Nowotny, H. and Taschwer, K. (eds) (1996) *The Sociology of Sciences*, 2 vols. Cheltenham: Elgar.

O'Brien, M. (1981) *The Politics of Reproduction*. London: Routledge and Kegan Paul.

O'Brien, M. (1989) *Reproducing the World: Essays in Feminist Theory.* Boulder, Col.: Westview Press.

Owen, T. (2006a) 'Towards a Post-Foucauldian Sociology of Ageing', in J.L. Powell and A. Wahidin (eds), *Foucault and Ageing*. New York: Nova Science Publishers.

Owen, T. (2006b) 'Genetic-social science and the study of Human Biotechnology', *Current Sociology*, 54(6): 897-917.

Owen, T. [2007a] 'After postmodernism: towards an evolutionary sociology', in J.L. Powell and T. Owen (eds) *Reconstructing Postmodernism: Critical Debates.* New York: Nova Science Publishers.

Owen, T. [2007b] 'Culture of crime control: through a post-Foucauldian lens', *The Internet Journal of Criminology [www.internetjournalofcriminology.com].*

Owen, T. and Powell J.L. (2006) '"Trust", professional power and social theory: lessons from a post-Foucauldian framework', *International Journal of Sociology and Social Policy*, 26(3/4): 110-120.

Office of Population, Consensus and Surveys (1993) Congenital malformation statistics. Her Majesty's Stationery Office, in R. Lynn (2001) *Eugenics: A Reassessment.* Westport, Connecticut: Praeger.

Opitz, J.M. (2000) 'The Geneticization of Western Civilisation: Blessing or Bane?', in P.R. Sloan (ed). *Controlling Our Destinies: Historical, Philosophical, Ethical, and Theological Perspectives on the Human Genome Project.* Notre Dame: Notre Dame Press.

Parry, S. (2003) 'The politics of cloning: mapping the rhetorical convergence of embryos and stem cells in parliamentary debates', *New Genetics and Society,* 22(2): 145-168.

Parsons, W. (1995) *Public Policy: An Introduction to the Theory and Practice of Policy Analysis.* Aldershot: Edward Elgar.

Paul, D. (1998) *The Politics of Heredity: Essays on Eugenics, Biomedicine, and The Nature-Nurture Debate.* Albany: State University of New York Press.

Peacocke, A.R. (2000) 'Relating Genetics to Theology on the Map of Scientific Knowledge', in P.R. Sloan (ed). *Controlling Our Destinies: Historical, Philosophical, Ethical, and Theological Perspectives on the Human Genome Project.* Notre Dame: Notre Dame Press.

Pebley, A.R. and Westaff, C.F. (1982) 'Women's sex preferences in the United States: 1970 to 1975', Demography, 19: 177-90, in R. Lynn (2001) *Eugenics: A Reassessment.* Westport, Connecticut: Praeger.

Pernick, M.S. (2000) 'Defining the Defective: Eugenics, Esthetics, and Mass Culture in Early Twentieth Century America', in P.R. Sloan (ed). *Controlling Our Destinies: Historical, Philosophical, Ethical, and Theological Perspectives on the Human Genome Project.* Notre Dame: Notre Dame Press.

Perrow, C. (1984) *Normal Accidents: Living with High Risk Technologies.* New York: Basic Books.

Peters, T. and Bennett, G.J. (2004) 'Science as saviour?' cloning, ethics and politics', *Dialog,* 43(3).

Peterson, A. and Bunton, R. (2002) *The New Genetics and the Public's Health.* London: Routledge.

Pickering, A. (1995a) 'Cyborg history and the World War II regime' *Perspectives on Science,* 3: 1-48.

Pickering, A. (1995b) *The Mangle of Practice: Time, Agency and Science.* Chicago: University of Chicago Press.

Pickering, A. (2001) *The Mangle of Practice: Time, Agency and Science.* 2nd edn. Chicago: University of Chicago Press.

Pierre, J. and Peters, B.G. (2000) *Governance, Politics and the State.* London: Macmillan.

Pinker, S. (1994) *The Language Instinct: The New Science of Language and Mind.* London: Penguin.

Pinker, S. (1995) *The Language Instinct.* London: Penguin.

Pinker, S. (1997) *How the Mind Works.* New York: W.W. Norton.

Pinker, S. (1999) *How the Mind Works* 2nd edn. London: Penguin.

Pinker, S. (2002) *The Blank Slate.* London: Penguin.

Pinker, S. and Rose, R. (21/01/1998) 'The Two Steves: Pinker vs Rose: a debate', *Edge,* available at http//www.edge.org/3rd_culture/pinker_rose/pinker_rose_p1.html.

Plomin, R. (1997) *Behavioural Genetics.* London: W.H. Freeman.

Pokorski, R.J. (1997) 'Insurance underwriting in the genetic era', *American Journal of Human Genetics*, 60: 205-216.

Pollock, K. (1988) 'On the nature of social stress: production of a modern mythology', *Social Science and Medicine*, 26(3): 381-392.

Powell, J. (1988) ''The Us and the Them': connecting Foucauldian and political economy insights into aging bodies'. *Paper presented to the British Sociological Association Annual Conference, University of Edinburgh.*

Powell, J.L. (2001) 'Social theory and the ageing body', *International Journal of Language, Society and Culture*, 8(2): 2-16.

Powell, J.L. and Biggs, S. (2000) 'Managing old age: the disciplinary web of power, surveillance and normalisation', *Journal of Aging and Identity,* 5, (1): 3-13.

Powell, J.L. and Biggs, S. (2001) 'Genealogies of old age and social welfare' in *Journal of Aging and Social Policy*, 12(2), 93-115.

Powell, J.L. and Biggs, S. (2004) 'Aging, self and illness in (Post) Modernity: A Foucauldian analysis', *International Journal of Sociology and Social Policy.*

Powell, J.L. and Owen, T. (2005) 'The bio-medical model and aging: towards an anti-reductionist model?', *International Journal of Sociology and Social Policy*, 25(9): 27-40.

Quilley, S. and Loyal, S. (2005) 'Eliasian sociology as a 'central theory' for the human sciences', *Current Sociology*, 53(5): 807-828.

Radcliffe Richards, J. (2001) *Human Nature After Darwin: A Philosophical Introduction.* London: Routledge.

Ratzinger, J. and Bovone, A. (1987) *Instruction on respect for human life in its origin and on the dignity of procreation.* Vatican City, Italy: Vatican Polyglot Press in R. Lynn (2001) *Eugenics: A Reassessment.* Westport, Connecticut; Praeger.

Rechsteiner, M. (1991) 'The Human Genome Project: misguided science policy?', *Trends in Biochemical Sciences,* 16: 453-59.

Renn, O. (1992) 'Concepts of risk: a classification', in S. Krimsky and D. Golding (eds) *Social Theories of Risk.* Westport, Connecticut: Praeger.

Revelli, A. et al (2003) *Biotechnology of Human Reproduction.* London: Taylor and Francis.

Rhodes, R.A.W. (1997) *Understanding Governance: Policy Networks, Governance, Reflexivity and Accountability.* Buckingham: Open University Press.

Rich, A. (1981) 'Disobedience is what NWSA is potentially about', *Women's Studies Quarterly*, 9(3): 4-5.

Richardson, J. (1996) 'Policy-making in the EU: interests, ideas, and garbage cans of primeval soup', in J. Richardson (ed), *European Union: Power and Policy Making.* London: Routledge.

Ridley, M. (1999) *Genome: The Autobiography of a Species in 23 Chapters.* London: Fourth Estate.

Ridley, M. (2003) *Nature Via Nurture: Genes, Experience and What Makes Us Human.* London: Fourth Estate.

Ritzer, G. (1990) 'Micro-macro linkages in sociology; applying a metatheoretical tool', in G. Ritzer (ed) *Frontiers of Social Theory; The New Syntheses.* New York: Columbia University Press.

Ritzer, G. (1992) 'Metatheorising in sociology: explaining the coming of age', in G. Ritzer (ed.) *Metatheorizing*. London: Sage.

Ritzer, G. (2000) *Modern Sociological Theory*. New York: McGraw-Hill.

Ritzer, G.and Ryan, J.M. [2007] 'Postmodern social theory and sociology: on symbolic exchange with a ''dead'' theory', in J.L.Powell and T.Owen [eds.] *Reconstructing Postmodernism: Critical Debates.* New York: Nova Science Publishers.

Roberts, D.E. (1996) 'Race and the new reproduction', *Hastings Law Journal,* 47: 935-49.

Rojek, C. and Turner, B. (2000) 'Decorative sociology: towards a critique of the cultural turn', *The Sociological Review*, 48(4): 629-48.

Rogers, L. (1999, July 4th) 'Disabled children will be a "sin", says scientist', Sunday Times (London), p.15, in R. Lynn (2001) *Eugenics: A Reassessment.* Westport, Connecticut: Praeger.

Rose, H. (2000) 'Colonising the social sciences?', in H. Rose and S. Rose, A*las Poor Darwin: Arguments Against Evolutionary Psychology.* London: Jonathan Cape.

Rose, H. and Rose, S. (eds) (2000) *Alas, Poor Darwin: Arguments Against Evolutionary Psychology.* London: Jonathan Cape.

Rose, S. (ed) (1982) *Towards a Liberatory Biology*. London-New York: Allison and Busby.

Rose, S. (1997) *Lifelines: Biology, Freedom, Determinism.* London: Penguin.

Rose, S. (ed) (1999) *From Brains to Consciousness: Essays on the New Science of the Mind.* London: Penguin.

Rose, S. (2000) 'Escaping evolutionary psychology', in H. Rose and S. Rose (eds) *Alas, Poor Darwin: Arguments Against Evolutionary Psychology.* London: Jonathan Cape.

Rose, S., Kamin, L.J. and Lewontin, R.C. (1984) *Not in Our Genes*. London: Pantheon.

Rose, N. and Miller, P. (1992) 'Political power beyond the state: problematics of government', *British Journal of Sociology*, 43(2): 173:205.

Royal College of Physicians (1989) 'Prenatal diagnosis and genetic screening'. London: Royal College of Physicians, in R. Lynn (2001) *Eugenics: A Reassessment.* Westport, Connecticut: Praeger.

Rubinsztein, D., Leggo, J., and Coles, R. et al (1996) 'Phenotypic characterization of individuals with 30-40 CAG repeats in the Huntington Disease (HD) gene reveals HD cases with 36 repeats and apparently normal elderly individuals with 36-39 repeats'. American Journal of Human Genetics 59:16-22, in P.R. Sloan (eds) (2000) *Controlling Our Destinies: Historical, Philosophical, Ethical, and Theological Perspectives on the Human Genome Project*. Notre Dame: Notre Dame Press.

Rubenstein, D.S., Thomasma, D.C., Schon, E.A., and Zinaman, M.J. [1995] 'Germ-line therapy to cure mitochondrial disease: protocol and ethics of in vitro ovum nuclear transplantation', *Cambridge Quarterly of Healthcare Ethics*, [4]:316-39.

Russett, B. and Starr, H. (1996) *World Politics: The Menu for Choice*. New York: Freeman and Company.

Russell, A. and Vogler, J. (2001) *The International Politics of Biotechnology: Investigating Global Futures*. Manchester: Manchester University Press.

Sahlins, M.D. (1972) *Stone Age Economics*, Chicago: Aldine-Atherton

Saussure, F. de ((1916)1974) *Course in General Linguistics*, trans W. Baskin. London: Collins.

Scambler, G. (2000) *Health and Social Change: A Critical Theory.* Buckingham: Open University Press.

Schaffner, K.F. (2000) 'Reductionism and determinism in human genetics: lessons from simple organisms', in P.R. Sloan (ed). *Controlling Our Destinies: Historical,*

Philosophical, Ethical, and Theological Perspectives on the Human Genome Project. Notre Dame: Notre Dame Press.

Schleiermacher, S. (1990) 'Racial hygiene and "Deliberate Parenthood": two sides of demographer Hans Harmsen's population policy', *Issues in Reproductive and Genetic Engineering: Journal of International Feminist Analysis*, 3(3): 201-10.

Scholes, R. (1974) *Structuralism in Literature*. New Haven, CT: Yale University Press.

Scruton, R. (1994) Modern Philosophy: An Introduction and Survey. London: Arrow.

Seidman, S. (1992) 'Postmodern social theory as narrative with a moral intent', in S.

Seidman and D. Wagner (eds) *Postmodernism and Social Theory*. Oxford: Blackwell.

Seidman, S. (1994) Contested Knowledge: Social Theory in the Postmodern Era. Oxford: Blackwell.

Shapin, S. (1982) 'History of science and its sociological reconstructions', *History of Science*, 20: 157-211.

Sharp, R. (1980) Knowledge, Ideology and the Politics of Schooling: Towards a Marxist Analysis of Schooling. London: Routledge.

Shaw, M.W. (1984) 'To be or not to be? That is the question', *American Journal of Human Genetics*, 36, 1-9, in R. Lynn (2001) *Eugenics: A Reassessment*. Westport, Connecticut: Praeger.

Shilling, C. (1993) *The Body and Social Theory*. London: Sage.

Shiva, V. (1988) *Staying Alive: Women, Ecology and Development*. London: Zed.

Sibeon, R. (1996) *Contemporary Sociology and Policy Analysis: The New Sociology of Public Policy*. London: Kogan Page and Tudor.

Sibeon, R. (1997a) 'Power agency/structure and micro-macro; an excursus in anti-reductionist sociology', Paper presented at *Annual Conference of the British Sociological Association, University of York, (April).*

Sibeon, R. (1997b) 'Anti-reductionist sociology and the study of postnational governance', *Paper presented at the Third Conference of the European Sociological Association*, University of Essex (September).

Sibeon, R. (1999) 'Anti-reductionist sociology', *Sociology*, 32, 2: 317-334.

Sibeon, R. (2001) 'The Reconstruction of sociological theory', *Paper presented at the 50th Anniversary Conference of the British Sociological Association, Manchester, 9-12 (April).*

Sibeon, R. (2004) *Rethinking Social Theory*. London: Sage.

Sibeon, R. [2007] 'An excursus in post-postmodern social science', in J.L.Powell and T.Owen [eds.] *Reconstructing Postmodernism: Critical Debates*. New York: Nova Science Publishers.

Singer, E. (1991) 'Public attitudes toward genetic testing', *Population Research and Policy Review*, 10: 235:55.

Sloane, P.R. (ed) et al (2000) *Controlling Our Destinies: Historical, Philosophical, Ethical, and Theological Perspectives on the Human Genome Project*. Notre Dame: Notre Dame Press.

Smart, B. [2007] '[Dis] interring postmodernism or a critique of the political economy of consumer choice', in J.L.Powell and T.Owen [eds.] *Reconstructing Postmodernism: Critical Debates*. New York: Nova Science Publishers.

Soja, E. (1989) *Postmodern Geographics: The Reassertion of Space in Critical Theory*. London: Verso.

Sokal, A. (1996) 'Transgressing the boundaries: towards a transformative hermeneutics of quantum gravity', *Social Text*, 14, (1-2): 217-252.

Sokal, A. and Bricmont, J. (1987) Impostures intellectuelles, Paris, Editions Odile Jacob (English trans. Sokal, A. and Bricmont, J. (1998) *Fashionable Nonsense: Postmodern Intellectuals' Abuse of Science*. New York: Picador.

Stones, R. (1996) *Sociological Reasoning: Towards a Past-Modern Sociology.*_London: Macmillan.

Sturgis, P., Cooper, H., and Fife-Schaw, C. (2005) 'Attitudes to biotechnology: estimating the opinions of a better-informed public', *New Genetics and Society*, 24(1) 4: 31-56.

Sulston, J. and Ferry, G. (2003) *The Common Thread: Science, Politics, Ethics and the Human Genome.* London: Corgi.

Suzuki, D. and Knudtson, P. (1990) 'Genetics: the clash between the new genetics and human values'. Cambridge M.A.: Harvard University Press, in R. Lynn (2001) *Eugenics: A Reassessment.* Westport, Connecticut: Praeger.

Swingewood, A. (2000) *A Short History of Sociological Thought.* London: Macmillan.

Symons, D. (1979) *The Evolution of Human Sexuality.* New York: Oxford University Press.

Symons, D. (1995) 'Beauty is in the adaptations of the beholder: the evolutionary psychology of human female sexual attractiveness', in P.R. Abramson and S.D. Pinkerton, *Sexual Nature, Sexual Culture.* Chicago: University of Chicago Press.

Tattersall, I. (1998) *Becoming Human.* New York: Harcourt Brace.

Tattersall, I. Email correspondence in Ridley, M. (2003) *Nature via Nurture: Genes, Experience and What Makes Us Human.* London: Fourth Estate.

Tawber, A.I. and Sarker, S. (1992), 'The Human Genome Project: has blind reductionism gone too far?', *Perspectives in Biology and Medicine,* 35: 220-35.

Taylor, S. (2006) *The Sociology of Emotion: An Introduction.* London: Palgrave.

Testart, J. (1995) 'The new eugenics and medicalised reproduction', Cambridge Quarterly of Health Care Ethics, 4, 304-12 in R. Lynn (2001) *Eugenics: A Reassessment.* Westport, Connecticut: Praeger.

Timson, J. (1997) 'Human genetics', The Galton Institute, edited by R.A. Peel, *Essays in the History of Eugenics:* proceedings of a conference organised by the Galton Institute, London, 1997.

Thain, M. and Hickman, M. (1995) *The Penguin Dictionary of Biology.* London: Penguin.

Tooby, J. (1999) 'The most testable concept in biology, Part 1', *HBES Newsletter* (Fall) (Available at: http://www.psych.ucsb.edu/research/cep/viewfall99.html).

Tooby, J. and Cosmides, L. (1990) 'The past explains the present: emotional adaptations and the structure of ancestral enviornments', *Ethnology and Sociobiology*, 11: 375-424.

Tooby, J. and Cosmides, L. (1992) 'The psychological foundations of culture', in J.H. Barkow, L. Cosmides and J. Tooby, *The Adapted Mind: Evolutionary Psychology and the Generation of Culture.* New York: Oxford University Press.

Tooby, J. and De Vore, I. (eds) (1987) 'The reconstruction of hominid behavioural evolution through strategic modelling', in *The Evolution of Human Behaviour: Primate Models.* Albany, New York: SUNY Press.

Touraine, A. (1981) *The Voice and the Eye: An Analysis of Social Movements.* Cambridge: Cambridge University Press.

Touraine, A. (1995) *Critique of Modernity.* Oxford: Blackwell.

Trivers, R. (1981) 'Sociobiology and politics', in E. White (ed) *Sociobiology and Human Politics*. Lexington, Mass., D.C.: Heath and Company.

Turner, B.S. and Rojek, C. (2001) *Society and Culture: Principles of Scarcity and Solidarity*. London: Sage.

Turner, B.S. [1990] 'Conclusion: peroration on ideology', in N.Abercrombie, S.Hill and B.S.Turner [eds.] *Dominant Ideologies*. London: Unwin Hyman.

Turner, L. [2004] 'Biotechnology as religion', *Nature Biotechnology*, 22 [6]: 604-659.

Verlinsky, Y. Pergament, E., and Strom, C. (1990) 'The preimplantation diagnosis of genetic diseases', Journal of In Vitro Fertilization and Embryo Transfer, 7, 1-5, in R. Lynn (2001) *Eugenics: A Reassessment*. Westport, Connecticut: Praeger.

Van Gelder, T. (1995) 'What might cognition be, if not computational?' *Journal of Philosophy*, 91 (7): 345-81.

Van Loon, J. (2002) *Risk and Technological Culture: Towards a Sociology of Violence*. London: Routledge.

Wade, N. (1998) 'It's a three-legged race to decipher the human genome', *New York Times* (23rd June): 811.

Wahidin, A. (2003) 'Reclaiming agency: managing ageing bodies in prison', in E. Tulle (ed) *Old Age and Human Agency*. New York: Nova Science Publishers.

Wainwright, D. and Calman, M. (2002) *Work Stress: The Making of a Modern Epidemic*. Buckingham: Open University Press.

Walby, S. (1990) *Theorising Patriarchy*. Oxford: Blackwell.

Walker, A. and Shipman, P. (1986) *The Wisdom of the Bones: In Search of the Human Origins*. London: Weidenfeld and Nicolson.

Wakayama, T., Perry, A.C., Zuccotti, M., Johnson, K.R., and Yanagimachi, R. (1998) 'Full-term development of mice from enucleated oocytes injected with cumulus cell nuclei', *Nature*, 394: 369-74, in R. Lynn (2001) *Eugenics: A Reassessment*. Westport, Connecticut: Praeger.

Warnock, M. (1987) 'Do human cells have rights?', *Bioethics*, 1: 8, in R. Lynn (2001) *Eugenics: A Reassessment*. Westport, Connecticut: Praeger.

Weatherall, D.J. (1991) The New Genetics and Clinical Practice. Oxford: Oxford University Press, in R. Lynn (2001) ibid.

Webster, A. [2005] 'Social science and a post-genomic future: alternative readings of genomic agency', *New Genetics and Society*, 24 [2]: 227-238.

Wertz, D.C. (1998) 'Eugenics is alive and well: A survey of genetic professionals around the world', *Science in Context*, 11: 493-510.

West, M. (2003) *The Immortal Cell: One Scientist's Quest to Solve the Mystery of Human Aging*. New York: Doubleday.

White, H.C. (1992) *Identity and Control: A Structural Theory of Social Action*. Princeton, N.J.: Princeton University Press.

Wilkie, T. (1994) *Perilous Knowledge: The Human Genome Project and Its Implications*. London: Faber and Faber Ltd.

Williams, S.J. (1996) 'The vicissitudes of embodiment across the chronic illness trajectory', *Body and Society*, 2 (2): 23-47.

Williams, S.J. (1998) ''Capitalising' on emotions? Rethinking the inequalities in health debate', *Sociology*, 32(1): 121-139.

Williams, S.J. (2003) 'Marrying the social and the biological? A rejoinder to Newton', *Sociological Review*, 51(4): 550-561.

Williams, S.J. and Bendelow, G. (1998) *The Lived Body: Sociological Themes Embodied Issues.* London: Routledge.

Williams-Jones, B., and Graham, J.E. (2003) 'Actor-network theory: a tool to support ethical analysis of commercial genetic testing', *New Genetics and Society*, 22(3): 271-296.

Wilkinson, R.G. (1996) *Unhealthy Societies: The Affliction of Inequality*. London: Routledge.

Wilkinson, R.G. (2000a) *Mind the Gap: Hierarchies, Health and Human Evolution*. London: Weidenfeld and Nicolson.

Wilkinson, R.G. (2000b) 'Deeper than 'neoliberalism': a reply to David Coburn', *Social Science and Medicine,* 51: 997-1000.

Wilmut, I., Schnieke, A., McWhis, J., Kind, A., and Campbell, K.H. (1997) 'Viable offspring derived from fetal and adult mammation cells', *Nature*, 390:27, in R. Lynn (2001) *Eugenics: A Reassessment*. Westport, Connecticut: Praeger.

World Health Organisation (1997, March 11). Press release on cloning, 20, in R. Lynn (ibid).

Ziman, J. (2001) *Real Science. What It Is, and What It Means*. Cambridge: Cambridge University Press.

Zimmerman, S. (1990) 'Industrial capitalism's "hostility to childbirth", "responsible childbearing" and eugenic reproductive politics in the first third of the 20th century', *Issues in Reproductive and Genetic Engineering: Journal of International Feminist Analysis,* 3(3): 191-200.

INDEX

A

D

E

F

G

H

I

J

K

L

M

N

Q

R

S

T

U

V

W